LEONARDO'S
MOUNTAIN OF
CLAMS AND THE
DIET OF WORMS

LEONARDO'S MOUNTAIN OF CLAMS AND THE DIET OF WORMS

ESSAYS ON NATURAL HISTORY

STEPHEN JAY GOULD

HARMONY BOOKS
New York

Published by Harmony Books, a division of Crown Publishers, Inc., 201 East 50th Street, New York, New York 10022. Member of the Crown Publishing Group.

Random House, Inc. New York, Toronto, London, Sydney, Auckland.
www.randomhouse.com
HARMONY and colophon are trademarks of Crown Publishers, Inc.

All of the essays contained in this work were previously published in *Natural History* magazine.

Printed in the United States of America

DESIGN BY LYNNE AMFT

Library of Congress Cataloging-in-Publication Data
Gould, Stephen Jay.
Leonardo's mountain of clams and the Diet of Worms : essays on
natural history / by Stephen Jay Gould. — 1st ed.
Includes bibliographical references and index.
1. Natural history.   2. Evolution (Biology).   I. Title.
QH81.G67323   1998
508—dc21   98-11500

ISBN 0-609-60141-5

10 9 8 7 6 5 4 3 2 1

First Edition

TO RAY SIEVER
AND TO THE MEMORY OF BERNIE KUMMEL,
two dear colleagues and friends
who nurtured (and protected)
a young pain in the ass
and helped him to become a scientist

CONTENTS

CONTENTS

LEONARDO'S
MOUNTAIN OF
CLAMS AND THE
DIET OF WORMS

PIECES OF EIGHT:

CONFESSION OF A

HUMANISTIC NATURALIST

I CAN EASILY UNDERSTAND WHY, FOR MOST NATURALISTS, THE HIGHEST form of beauty, inspiration, and moral value might be imputed to increasingly rare patches of true wilderness—that is, to parcels of nature devoid of any human presence, either in current person or by previous incursion. When we recognize that all but the last geological eyeblink of life's history evolved in competence and fascination (but to whose notice?) before humans intruded upon the scene—and when we acknowledge that most of our substantial incursions cannot be viewed as fortunate either for local organisms or environments—why should we not glory in bits of space that have perpetuated a 4.5-billion-year tradition of noninterference by any self-conscious agency? (As I do not wish to engage the theological dimensions of the last sentence, I will restrict my meaning to overt "footprints" of undeniable physical presence.)

I do have a confession to make in this context. My odd attitude may arise only from the happenstance of my birth and happy childhood in New York City, when safe subways cost a nickel, museums were free, and the Yankees, led by Joe DiMaggio, ruled the world. Wordsworth's wisdom cannot be gainsaid. Childhood's sense of wonder cannot be sustained in the same manner through life, but the child is father to the man. So childhood's "splendor in the grass" and "glory in the flower" must set a lifelong prototype for aesthetic wonder. And my early epiphanic moments included the view of Lower Manhattan's buildings at sunset, seen from the magnificent walkway in the center of the Brooklyn Bridge; the growing tip of Manhattan as the Staten Island Ferry (also only a nickel) passes the Statue of Liberty and heads for the Battery; the lobbies of the Woolworth and Chrysler buildings (each, in turn and temporarily, the tallest skyscraper in the world); and the building line of the surrounding city, seen in winter from the middle of Central Park through bare tree branches.

I am not speaking here, by absurd dichotomy, of city versus wilderness, with a personal preference for the former based on accidents of upbringing. Rather, the dichotomy itself has no meaning, if only because "pure" examples of either extreme scarcely exist when plastic flotsam pervades the seas, and twisted jetsam washes up on the beaches of every isolated and uninhabited Pacific island; and when almost every spot perceived with rapture as "virgin" wilderness (at least here in northeastern America) really represents old farmland reclaimed by new forest. No satanic "purity" marks the other end either, except in science fiction scenarios. We do not build cities without parks, streets without trees, homes without gardens. At a bare minimum, bits of nature's diversity still burst through, if only as rats by the garbage piles, cockroaches in the kitchen, mushrooms through the pavement, weeds galore in the lot, and bacteria everywhere—to cite all major kingdoms of life in the big city.

For whatever reasons of childhood's happenstances and gifts of temperament, I am a humanist at heart, and I love, best of all, the sensitive and intelligent conjunction of art and nature—not the domination of one by the other. We want, in our wondrously diverse world, a full spectrum of interactions from near wilderness to near artificiality, but I will seek my own aesthetic optimum right in the middle, where human activity has tweaked or shaped a landscape, but with such respect and integration that a first glance may detect no fault line, no obvious partitioning: the wooded hillslope adjoining Kiyomizudera in Kyoto, where the gorgeous scene looks so perfectly "rustic" and untouched until you realize that every tree has been selected, pruned, and trained; the genius of Olmsted's big city parks, with their sculpted diversity of "natural" landscapes crisscrossed by a respectful system of constructed pathways, built of local stones artificially rusticated if necessary; the smooth transition between a Chinese "scholar's rock" (selected for calming contemplation based on the fortune of naturally formed beauty, but usually sculpted a bit to enhance the appearance), and the wooden stand expressly carved to accommodate every random bump and crevice of the stone above; and the Hopi pueblo towns, built of local rocks as a layer on the tops of mesas made of horizontal strata, so that the town, from a distance, can hardly be distinguished from the natural layers below, a village marked as a human construction only by vertical ladders protruding from the tops of kivas.

I even believe—though I would not push the point, for the concept can too easily cede to human arrogance and a discounting of natural forms— that intelligent reconstruction can "improve" upon natural design (though only by the criterion of human aesthetic preference, the most parochial of all possible judgments). I do ally myself with the most famous quatrain of Omar Khayyám's *Rubáiyát* (in FitzGerald's Victorian version), a passage usually misinterpreted today because the subjunctive mood has virtually disappeared from modern English:

> *A Book of Verses underneath the Bough,*
> *A Jug of Wine, a Loaf of Bread—and Thou*
> *Beside me singing in the Wilderness—*
> *Oh, Wilderness were Paradise enow!*

That is, if you would join me in the wilderness, and we could share good reading, food, drink (and perhaps more), then even the ugly, scary, untamed forest would become a paradise, literally a lovely *enclosed and cultivated garden*. (The old subjunctive of the last line must be read: "Even wilderness *would be* close enough to paradise" if you and all the accoutrements would join me there.) After all, in many cultures, wilderness (with an etymology of "wild beast") denotes fear and foreignness, while human cultivation tames a landscape to beauty and peace of soul. (I also love the old legend—maybe it's even true—that Eugene O'Neill changed Omar's last line to "Ah, Wilderness!" so that the title for his marvelous coming-of-age play would appear first in *The New York Times*'s alphabetical list of Broadway shows.)

I make this humanistic confession (or profession, really) because I have tried, in the prefaces to each of my essay volumes (this is the eighth in a series that will reach ten before the millennium calls a halt), to figure out how the present effort differs from (and, I hope, builds upon) the varying themes of preceding books. I began with emphasis on evolutionary basics, proceeded to evolutionary implications, social and philosophical usages, the interaction of predictive rules with contingent history to form the unique and surprising patterns of life's history, and the interaction of human history with natural environments.

This eighth volume, as usual, includes all these themes, but differs in emphasis primarily in my own increasing comfort with my unconventional approach to "natural history" writing, as outlined above. If any overarching theme pervades this body of writing (now standing at 270 successive monthly essays), I suppose that a groping effort toward the formulation of a humanistic natural history must unite the disparity. I think that I have been reluctant to recognize, address, or even admit this feature, either to myself or to my readers, because such an approach does contravene a deep (and usually unstated) convention in writing about nature. We are supposed to love nature for itself, and we are, therefore, presumably charged with the task of characterizing and interpreting nature (as she is) so that interested people with less expertise can learn new information and draw appropriate messages, both factual and ethical. Well, I do love nature—as fiercely as anyone who has ever taken up a pen in her service. But I am even more fascinated by the complex level of analysis just above and beyond (and I do mean "abstracted from," not "better than")—that is, the history of how humans have learned to study and understand nature. I am primarily a "humanistic naturalist" in this crucial sense.

Of course I yearn for answers to all the puzzles, great and small, that build the order (and wondrous disorder) of nature "out there"—an order that our intellectual ancestors could only read (understandably) as a proof of God's existence and benevolent intent. And I am convinced that such answers exist, if only to be seen "through a glass darkly," given the necessary interposition of human history, sociology, and psychology between the "real" world, and any abstractions of disembodied logic that might manipulate and order our observations. (In this sense, no practicing scientist can be a pure "relativist," although I trust the more sophisticated and self-analytical among us know that "pure" observation, "unsullied" by human foibles and preferences, can only rank as idealized legend.)

But I prefer to emphasize the interaction of this outside world with something unique in the history of life on Earth—the struggle of a conscious and questioning agent to understand the whys and wherefores, and to integrate this knowledge with the meaning of its own existence. That is, I am enthused by nature's constitution, but even more fascinated by trying to grasp how an odd and excessively fragile instrument—the human mind—comes to know this world outside, and how the contingent history of the human body, personality, and society impacts the pathways to this knowledge.

A map of the roadblocks—imposed by the evolutionary limitations of an instrument clearly not designed for this style of inquiry, and then joined with the improbable and unrepeatable contingencies that built our modern technological society—holds just as much interest as an accurate map of nature's geography. Moreover, a humanistic focus on how we know about nature—rather than an "objective" account, unattainable in any case, of how nature "is"—gives an essayist a "whole 'nother" level of juicy material, for we lose nothing of the primary topic, the world as we find it, and gain all the foibles and fascination of *how* we find it so.

As another benefit of this humanistic focus, we acquire a surprising source of rich and apparently limitless novelty from the primary documents of great thinkers throughout our history. But why should any nuggets, or even flakes, be left for intellectual miners in such terrain? Hasn't the *Origin of Species* been read untold millions of times? Hasn't every paragraph been subjected to overt scholarly scrutiny and exegesis?

Let me share a secret rooted in general human foibles, and in the faint tinge of anti-intellectualism that has always pervaded American culture. Very few people, including authors willing to commit to paper, ever really read primary sources—certainly not in necessary depth and completion, and often not at all. Nothing new here, but this shortcutting propensity of the ages has been abetted in our "journalistic" era by a lamentable tendency to call experts, rather than to read and ponder—yet another guarantee of authorial passivity before secondary sources, rather than active dialogue, or communion by study, with the great thinkers of our past.

I stress this point primarily for a practical, even an ethical, reason, and not merely to vent my spleen. When writers close themselves off to the documents of scholarship, and rely only on seeing or asking, they become conduits and sieves rather than thinkers. When, on the other hand, you study the great works of predecessors engaged in the same struggle, you enter a dialogue with human history and the rich variety of our intellectual traditions. You insert yourself, and your own organizing powers, into this history—and you become an active agent, not merely a "reporter." Then, and only then, can you become an original contributor, even a discoverer, and not only a mouthpiece.

What could be more democratic than the principle that nuggets of real discovery abound in primary sources, located in such accessible places as major university and city libraries, for those willing to do the work and develop the skills. (And there's the rub. I do, of course, acknowledge the impediment for most Americans that many of these works, representing the ecumenical range of international scholarship, have never been translated into English—a fact that should be a spur to study, and not a barrier.) Good anatomists have told me that novel and important observations can still be made by dissecting a common frog, despite millions of prior efforts spanning several centuries. I can attest that all major documents of science remain chock-full of distinctive and illuminating novelty, if only people will study them—in full and in the original editions. Why would anyone *not* yearn to read these works; not hunger for the opportunity? What a thrill, whatever the outcome in personal enlightenment, to thus engage the greatest thinkers and doers of our past, to thumb the pages of their own printings, to speculate about past readers who pondered the same copies with the differing presuppositions of other centuries, as the candle of nighttime illuminated their silent labor.

Of the six parts in this humanist's natural history of evolutionary essays, the first four—on art and science, mini-biographies, human prehistory with emphasis on paleolithic cave art, and human history from a naturalist's standpoint—emphasize our side, though several focus on particular organisms, as in chapter 9 on giant deer ("Irish elks") painted on cave walls, chapter 11 on Bahamian land snails for a fable about Columbus, and chapter 12 on the dodo's fate, made even sadder by human insult added to the ultimate injury of extirpation. The essays of the last two sections—on evolutionary theory, and on perspectives of other organisms—focus on the nonhuman side (again with such exceptions, as chapter 14 on papal statements about evolution, chapter 15 on the contrast of Robert Boyle and Charles Darwin on natural design, and chapter 18 on Percival Lowell versus Alfred Russel Wallace on Martian canals and the true domination of earthly life by bacteria.)

All these essays are grounded in a precious paradox that has defined the best of the genre ever since Montaigne: intimate and accurate detail—the foundation of most good essays—serves as a source of delight in itself, and also as a springboard to discourse about generalities of broadest scope. I would never dare to take on "the nature of truth" by frontal assault and abstract generalization—for fear of becoming an empty, tendentious buffoon, pontificating about the unanswerable and undefinable. But the subject must rivet us, and we can legitimately "sneak up on" (and even genuinely illuminate) this great issue by discussing how Darwin and his creationist American soulmate Dana constructed alternative taxonomies for toothed birds that should not have existed under previous concepts of reality, but had just been discovered as fossils (chapter 5). Similarly, if I tackled "the nature of tolerance" head-on, naked of intriguing and specific illustration, I would sound like a vain preacher crying in the wilderness (negative definition!). But if I confess some childhood humor in juxtaposing, for alliteration as well as content, the Diet of Worms with the Defenestration of Prague (chapter 13), then a seemingly superficial, even ridiculous, union wins legitimacy for joint illustration, and provides fair access to factual and moral dimensions of the general topic.

These essays probe, arrange, join, and parry the details within a diverse forest of data, located both in nature and in the documents of human struggle—all to access an inherently confusing but infinitely compelling world. As I survey the contents of this eighth volume, I find that I have followed four primary strategies to promote these details into coherent frameworks with sufficient generality to incite an essay.

1. In some cases, an intense study of original sources yields genuine discovery, despite the paradox that materials for a solution have always been patent. The story of nonuse for the giraffe's neck by early evolutionists had not been documented before (chapter 16), and surprising absences often reveal as much as unrecognized presences. I located a new dimension, largely in favor of the "vanquished" Owen and not the "victor" Huxley, in the great hippocampus debate that animated evolutionary discussion in the 1860s (chapter 6). Dana's important theory of cephalization, and its link with his natural theology (in interesting contrast with Darwin's developing

alternative), has never been elucidated, in part because Dana scattered his views through so many short and technical papers (chapter 5).

But I am, I confess, most proud of the opening title essay on Leonardo's paleontology. The excellence and prominence of his observations on fossils have been recognized—and dutifully honored in all accounts, popular, textbook, and technical—for more than a century, since the full publication of his private notebooks in the 1880s. But no one had identified the special reasons (based on his own, and largely medieval, views of the earth as analogous to a living body) for his intense focus on fossils, and for the placement of his statements in a codex largely devoted to the nature of water. So these wonderful observations had stood out, disembodied from context, and misinterpreted as the weird anachronisms of a transcendent and largely unfathomable genius. But the full document of the Leicester Codex sets the proper context, when read in its entirety and understood by the physics of Leonardo's own time.

2. In most cases, I do not report observations never made before, but try to place unfamiliar (or even well-known) items into a novel context by juxtaposition with other subjects not previously viewed as related—invariably in the service of illuminating a general point about the practice of science, the structure of nature, or the construction of knowledge. In reviewing the essays for this volume (not planned as an ensemble when first written, but collected from my monthly series for *Natural History* magazine), I noticed that I had most often made such a juxtaposition by the minimal method of pairing, or contrast between two—perhaps a general mode of operation for the human mind, at least according to several prominent schools of research (discussed here in the context of paleolithic cave art in chapter 8). For example, all the essays in part 2 on mini-biographies, although focusing on one previously unappreciated or misunderstood character, interpret their subject by his contrast with a standard figure—Linnaeus and the eighteenth-century English Jewish naturalist Mendes da Costa (chapter 4),

James D. Dana and his British soulmate Darwin (chapter 5), Richard Owen versus T. H. Huxley (chapter 6), and the tragic Russian genius Vladimir Kovalevsky (and his equally tragic and more brilliant wife, Sophia, one of the greatest mathematicians of the nineteenth century) with Darwin on the potential of error to illuminate scientific truth (chapter 7).

Many other essays also pursue this strategy of illumination by paired contrast, with novelty in the joining: Boyle and Darwin on natural theology and evolution (chapter 15); Percival Lowell versus Alfred Russel Wallace on the canals of Mars and the uniqueness of life (chapter 18); sloths and vultures as prototypes for traits that we, in our parochial and irrelevant way, judge as negative but yearn to understand (chapter 20); the Diet of Worms and the Defenestration of Prague as events of European history, related by more than their shared initial *D* and funny names (chapter 13); the Abbé Breuil and André Leroi-Gourhan for two sequential and maximally contrasting (but strangely similar) theories about the genesis of cave art (chapter 8); the great artist Turner and the prime engineer Brunel on the similarity of art and science (chapter 2); a forgotten theory about the origin of vertebrates with stunning new data to validate an even older view, all as an entrée to the subject of major evolutionary transitions and the prejudices that impede our understanding of this topic (chapter 17); the dodo of Mauritius and the first New World victims of Western genocide (chapter 12); and the striking difference between two popes in their common willingness to support the factual truth of evolution (chapter 14).

3. If my second category works by joining disparate details, a third strategy operates by careful excavation—elucidation by digging rather than elucidation by joining. As the mineshaft widens and deepens, one may reach a richness of detail justifying promotion to an essay because the requisite generality has been attained by one of two routes: (1) By casting a truly novel, or at least sufficiently different, light on an old subject, so that readers become willing to devote renewed interest, and may even obtain some

provocative insight (Darwin always wrote to his creationist friends that he dared not expect to change their minds, but did hope to "stagger" them a bit)—as when intricate details of the life cycle of the maximally "degenerate" parasite *Sacculina* suggest new attention to the fallacies of evolutionary progress (chapter 19), and when the subtle (and almost entirely unreported) distinctions in the affirmation of evolution by two very different popes (Pius XII and John Paul II) illuminate the old and overly discussed issue of proper relationships between science and religion (chapter 14). (2) By gaining the "right" to address a large and general issue through the new perspective of previously unapplied detail (as in the examples of chapters 5 and 13, previously discussed, and chapter 10 on the relevance of new data about the multiplicity of human species until 30,000–40,000 years ago and the consequent oddity of our current status as a single species spread throughout the globe) for a discussion of predictability versus historical contingency in the evolution of self-conscious life on Earth.

4. "Promotion" to an essay may depend upon the coalescence of details into a general theme worthy of report, but sometimes those details, all by themselves, become arresting enough to merit treatment entirely for their own value (and then I will confess to using the emerging generality as an excuse for almost baroque attention to the details). I do value the theme eventually addressed, but don't you adore, entirely for their own sake as stories, the four tales of conventional prey that devour their predators (chapter 21), or the excruciatingly intricate and beautiful details of the bizarrely complex life cycle of the barnacle parasite, the "root-head" *Sacculina* (chapter 19)? And, as my personal favorite (and here I do rest my case), how could anyone but a dolt not be moved by the fact that we know about the giant deer's hump only because paleolithic cave painters left us a record—and that no other even potential source of evidence exists (chapter 9). I tell this story within a perfectly valid and sufficiently interesting context of discourse on biological adaptation as a general evolutionary principle, but don't you thrill

to the notion of this kind of gift provided by such distant forebears; and aren't you riveted by the details of these rare images, and the story of their discovery and recognition?

The foregoing discussion accounts for all individual bits in this eighth piece of my series. But just as the "two bits" of legend represented a cut from a totality called a "piece of eight,"[1] my bits have no coherence or valid generality without an overarching rationale or coordinating theme to make them whole. I pay my homage to evolution in the preface to every volume of this series, and will now do so again. Of all general themes in science, no other could be so rich, so deep, so fascinating in extension, or so troubling (to our deepest hopes and prejudices) in implication. Therefore, for an essayist in need of a ligature for disparate thoughts and subjects, no binder could possibly be more appropriate—in fascination and legitimacy—than evolution, the concept that inspired the great biologist Theodosius Dobzhansky to remark, in one of the most widely quoted statements of twentieth-century science, that "nothing in biology makes sense except in the light of evolution."

Moreover, and finally, with this series' emphasis on a humanistic natural history—an account of evolution that focuses as much on how we come to know and understand this great principle as on how such a process shapes the history of life—we encounter an endless recursion that provides even greater scope and interest to the subject. The wondrously peculiar human brain arose as a product of evolution, replete with odd (and often misleading) modes of reasoning originally developed for other purposes, or for no explicit purpose at all. This brain then discovers the central truth of evolution, but also constructs human cultures and societies, replete with hopes and prejudices that predispose us toward rejecting many modes and implications

1. The etymology is much disputed, but I will follow John Ciardi's *Browser's Dictionary* (Harper & Row, 1980) for the conventional story that American colonials (in the absence of an official mint before we became a nation) used the coins of several countries for change. The Spanish silver "piece of eight" (so called because the coin bore a large number 8 to signify its value as eight *reals*) was often cut into pieces, called "bits." Since the *real* was worth about 12½ cents, two bits became an American quarter, four bits a half-dollar, and so on—in terminology still used today.

of the very process that created us. And thus, in a kind of almost cosmically wicked recursion, evolution builds the brain, and the brain invents both the culture that must face evolution and the modes of reasoning that might elucidate the process of its own creation. Round and round we go—into a whorl that may be endless and eternal, yet seems to feature some form of increasing understanding in all the gyrations that, at the very least, give us topics for essays and, at best, provide some insight into the nature of our being.

I
ART AND SCIENCE

1

THE UPWARDLY MOBILE FOSSILS OF LEONARDO'S LIVING EARTH

MORGAN DESCRIBES HIS DESPAIR AS THEIR CAPTORS STRING UP KING Arthur for a hanging: "They were blindfolding him! I was paralysed; I couldn't move, I was choking, my tongue was petrified . . . They led him under the rope." But, in the best cliff-hanging traditions, and at the last conceivable instant, Sir Lancelot comes to the rescue with five hundred knights—all riding bicycles. "Lord, how the plumes streamed, how the sun flamed and flashed from the endless procession of webby wheels! I waved my right arm as Lancelot swept in. I tore away noose and bandage, and shouted: 'On your knees, every rascal of you, and salute the king! Who fails shall sup in hell to-night!'"

I am not citing either Monty Python or *Saturday Night Live,* and I didn't mix up my genders in the first sentence. The speaker is not Morgan le Fay (who, no doubt, would have devised a magical, rather than a technological, solution to the same predicament), but Hank Morgan, the

Connecticut Yankee in King Arthur's court, and the hero in Mark Twain's satirical novel of the same name. Morgan, transported from nineteenth-century Hartford, wreaks mayhem in sixth-century Camelot by introducing all manner of "modern" conveniences, including tobacco, telephones, baseball—and bicycles.

As a literary or artistic device, anachronism exerts a powerful hold upon us, and has been a staple of all genres from the highest philosophy to the lowest comedy—as Jesus is crucified in a corporate boardroom by Dali, condemned at his Second Coming by Dostoyevsky's Grand Inquisitor, but only offered a half-price discount (as he changes to modern dress) by the Italian barber or the Jewish tailor of various ethnic jokes, now deemed tasteless and untellable.

Anachronism works this eerie and potent effect, I suppose, because we use the known temporal sequence of our history as a primary device for imposing order upon a confusing world. And when "the time is out of joint. O cursèd spite," we really do get discombobulated. We also know that correction of a perceived time warp cannot be achieved so easily in real life as in magical fiction (where Merlin can put Hank Morgan to sleep for 1,300 years, or Dracula can be dispatched with a wooden stake driven into the right spot). We regard Hamlet's blithe confidence as a mark of his madness when he completes his rhyming couplet with the Shakespearean equivalent of "no sweat" or "hakuna matata": ". . . That ever I was born to set it right!"

Science, for reasons partly mythical, but also partly accurate and honorable, presents itself as the most linear and chronologically well ordered of all disciplines. If science, working by fruitful and largely unchanging methods of reason, observation, and experimentation, develops progressively more accurate accounts of the natural world, then history provides a time line defined by ever-expanding success. In such a simple linear ordering, mediated by a single principle of advancing knowledge, any pronounced anachronism must strike us as especially peculiar—and subject to diametrically opposite judgment depending upon the direction of warp. An ancient view maintained in the present strikes us as risible and absurd—the creationist who wants to compress the history of life into the few thousand years of a literal biblical chronology, or the few serious members of the Flat Earth Society. But a "modern" truth, espoused out of time by a scholar in the distant past, fills us with awe, and may even seem close to miraculous.

A person consistently ahead of his time—a real-life Hank Morgan who could present a six-shooter to Julius Caesar, or explain the theory of natural selection to Saint Thomas Aquinas—can only evoke a metaphorical comparison with a spaceman from a more advanced universe, or a genuine angel from the realms of glory. In the entire history of science, no man seems so well qualified for such a designation as Leonardo da Vinci, who died in 1519, but filled his private notebooks with the principles of aeronautics, the mental invention of flying machines and submarines, and a correct explanation for the nature of fossils that professional science would not develop until the end of the eighteenth century. Did he have a private line across the centuries to Einstein, or even to God Himself?

I must confess that I share, with so many others, a lifelong fascination for this man. I was not a particularly intellectual child; I played stickball every afternoon and read little beyond comic books and school assignments. But Leonardo captured my imagination. I asked, at age ten or so, for a book about his life and work, probably the only intellectual gift that I ever overtly requested from my parents. As an undergraduate geology major, I bought the two-volume Dover paperback edition of Leonardo's notebooks (a reprint of the 1883 compilation by Jean Paul Richter) because I had read some of his observations on fossils in the Leicester Codex,[1] and had been

---

1. What goes 'round, comes 'round—as Leonardo must have said somewhere. This Leicester Codex, one of Leonardo's most important notebooks, filled largely with commentary on the nature and use of water, first came to light in the 1690s, when Giuseppi Ghezzi found the document in a chest of manuscripts in Rome. In 1717, Thomas Coke, late Lord Leicester (hence the codex's name), purchased the notebook, which remained in his family until Armand Hammer bought it in 1980—and renamed it, in Trumpian fashion, Codex Hammer. With enormous fanfare (at enormous profit), Christie's auctioned this notebook on November 11, 1994, where America's Bill Gates outbid several European governments, and bought the manuscript for more money than I can count. Gates, to his credit, restored the original name, and has favored public exhibition of the document—including a show at the American Museum of Natural History in

stunned not only by their accuracy, but also by their clear statement of paleo-ecological principles not clearly codified before our century, and still serving as a basis for modern studies.

Leonardo remains, in many ways, a frustrating and shadowy figure. He painted only about a dozen authenticated works, but these include two of the most famous images in our culture, the Mona Lisa (in the Louvre) and the *Last Supper* (a crumbling fresco in Milan). He published nothing in his lifetime, despite numerous and exuberant plans, though several thousand fascinating pages of manuscript have survived, probably representing only about a quarter of his total output. He did not hide his light under a bushel and was, in life, probably the most celebrated intellectual in Europe. Dukes and kings reveled in his conversation and his plans for war machines and irrigation projects. He served under the generous patronage of Europe's most powerful rulers, including Ludovico il Moro of Milan, the infamous Cesare Borgia, and King Francis I of France.

Leonardo's notebooks did not become generally known until the late eighteenth century, and were not published (and then only in fragmentary and occasional form) until the nineteenth century. Thus, he occupies the unique and peculiar role of a "private spaceman"—a thinker of preeminent originality, but whose unknown works exerted no influence at all upon the developing history of science (for nearly all his great insights had been redis-covered independently before his notebooks came to light).[2]

1996, where I finally saw this icon of my dreams and admiration, and where I developed the ideas for this essay. The Leicester Codex is the only manuscript of Leonardo's now residing in America.

2. An air of impenetrability continues to surround Leonardo. A scholar must still struggle to obtain a complete translation of any document like the Leicester Codex. The Richter edition of Leonardo's notebooks is maddeningly fragmentary, and the individual passages of the codices are broken apart and rearranged by subject. (Thus, you can find Leonardo's statements about water under a common heading, as abstracted from all his notebooks, but you cannot put together the text of the Leicester Codex—admittedly a hodgepodge and miscellany, but scholars do need to trace sequential jottings, however motley the apparent medley, for Leonardo often made odd jux-tapositions for interesting reasons.) The other major edition of Leonardo's notebooks—Edward MacCurdy's compilation of 1939, and my source of quotation for this essay—is far more adequate

The Upwardly Mobile Fossils

The overwhelmingly prevailing weight of public commentary about Leonardo continues to view him as Western culture's primary example of a "spaceman," that is, as a genius so transcendent that he could reach, in his own fifteenth century, conclusions that the rest of science, plodding forward in its linear march to truth, would not ascertain for several hundred years. Leonardo stood alone and above, we are told over and over again, because he combined his unparalleled genius with a thoroughly modern methodology based on close observation and clever experiment. He could therefore overcome the ignorance and lingering sterile scholasticism of his own times.

For example, the "Introductory Note" in the official catalog for a recent exhibition of the Leicester Codex in New York summarizes the basis of Leonardo's success in these words: "In it [the codex] we can begin to see how he combined almost superhuman powers of observation with an understanding of the importance of experimentation. The results were inspired insights into the workings of nature that match his artistic achievements." When such conventional sources acknowledge the persisting medieval character of many Leonardian pronouncements, they almost always view this context as a pure impediment to be overcome by observation and experiment, not as a matrix that might have been useful to Leonardo, or might help us to understand his beliefs and conclusions. For example, the closing passage of the long *Encyclopaedia Britannica* article on Leonardo states:

(and nearly complete for the Leicester Codex), although also broken up by topic. I must confess to a wry amusement (which might have blossomed to near fury if I had a different temperament) at the recent exhibit of the Leicester Codex in New York's Museum of Natural History. Visitors could see all the original pages and buy a beautiful catalog with each page reproduced in full facsimile. But no printed translation could be found anywhere, and the catalog only provided a pitifully scrappy summary of each page. You could purchase a CD-ROM with the full text (as Bill Gates showed his true commitment!), but most homes don't have a machine for playback, and the version that I tried to use couldn't even put a full line, with Leonardo's marginal annotations, on the screen at once. Moreover, a scholar can't work with only one part of a text on a screen at a time. You have to be able to compare passages from several pages at once—as you can do with an old-fashioned book. I almost felt as though our modern age of the passive sound bite—the attitude of "we know what little bit you need"—had launched a conspiracy against scholarship to keep Leonardo hidden. I do love to consult original sources in their original languages, but my skills (and patience) do not extend to long bouts of reading medieval Italian in a mirror!

"Leonardo approached this vast realm of nature to probe its secrets . . . The knowledge thus won was still bound up with medieval Scholastic conceptions, but the results of his research were among the first great achievements of the thinking of the new age because they were based on the principle of experience."

I think that this conventional view could not be more wrong in its general approach to the history of knowledge, or more stultifying for our quest to understand this most fascinating man of our intellectual past. Leonardo did make wonderful observations. He did often anticipate conclusions that public science would not reach for another two or three centuries. But he was neither a spaceman nor an angel—and we will never understand him if we insist on reading him as Hank Morgan, a man truly out of time, a modernist among the Medici, a futurist in the court of Francis the First.

Leonardo operated in the context of his time. He used his basically medieval and Renaissance concept of the universe to pose the great questions, and to organize the subjects and phenomena, that would generate his phenomenal originality. If we do not chronicle, *and respect,* the medieval sources and character of Leonardo's thought, we will never understand him or truly appreciate his transforming ideas. All great science, indeed all fruitful thinking, must occur in a social and intellectual context—and contexts are just as likely to promote insight as to constrain thought. History does not unfold along a line of progress, and the past was not just a bad old time to be superseded and rejected for its inevitable antiquity.

In this essay, I will try to illustrate the centrality of Leonardo's largely medieval context by analyzing his remarkable paleontological observations in the Leicester Codex. I will begin by acknowledging their truly prescient character, but will then raise two questions that expose the early-sixteenth-century context of Leonardo's inquiry: first, "What alternative account of fossils was Leonardo trying to disprove by making his observations?" and, second, "What theory of the earth was Leonardo trying to support with his findings?" Leonardo did not make his observations to win the praises of future generations; he studied fossils to probe these two questions of his own time—and his answers could not be more deeply embedded in a "hot topic" of his own century that we would now mock and dismiss as hopelessly antiquated. Thus, we cannot understand Leonardo's paleontology when we only marvel at his empirical accuracy and ignore the reasons for his inquiry.

Yes indeed, a thousand times yes, Leonardo's observations are often stunningly accurate—as experts have always said, and for the reasons conventionally stated. Moreover, their degree of detail, and their centrality to the basic rules of modern paleoecological analysis, only enhance the impression of authorship by a Victorian geologist somehow trapped in the early sixteenth century. But let me stop marveling and start listing a small sample!

1. Leonardo recognized the temporal and historical nature of horizontal strata by correlating the same layers across the two sides of river valleys:

> How the rivers have all sawn through and divided the members of the great Alps one from another; and this is revealed by the arrangement of the stratified rocks, in which from the summit of the mountain down to the river one sees the strata on the one side of the river corresponding with those on the other.

(All quotes, unless otherwise stated, come from the Leicester Codex as presented in the MacCurdy translation of Leonardo's notebooks.)

2. He observed that rivers deposit large, angular rocks near their sources in high mountains, and that transported blocks are progressively worn down in size, and rounded in shape, until sluggish rivers deposit gravel, and eventually fine clay, near their mouths. (I learned this rule as principle number one on day number one in my college course in beginning geology.)

> When a river flows out from among mountains it deposits a great quantity of large stones . . . And these stones still retain some part of their angles and sides; and as it proceeds on its course it carries with it the lesser stones with angles more worn away, and so the large

> stones become smaller; and farther on it deposits first coarse and then fine gravel . . . until at last the sand becomes so fine as to seem almost like water . . . and this is the white earth that is used for making jugs.

3. The presence of fossils in several superposed layers proves their deposition at different and sequential times.

4. The tracks and trails of marine organisms are often preserved on bedding planes of strata: "How between the various layers of the stone are still to be found the tracks of the worms which crawled about upon them when it was not yet dry."

5. If both valves of a clam remain together in a fossil deposit, the animal must have been buried where it lived, for any extensive transport by currents after death must disarticulate the valves, which are not cemented together in life, but only hinged by an organic ligament that quickly decays after death. (This principle of inferring transport by noting whether fossil clams retain both valves persists as a primary rule of thumb for everyday paleoecological analysis. I doubt that any pre-nineteenth-century geologist mentioned this observation in more than a casual manner, while Leonardo regarded the argument as central. This observation first inspired my undergraduate awe for Leonardo, for I had just learned the rule in class and had thought, "How clever; how modern.")

> And we find the oysters together in very large families, among which some may be seen with their shells still joined together, which serves to indicate that they were left there by the sea and that they were still living.

At another site, on the other hand, Leonardo inferred extensive transport after death:

> In such a locality there was a sea beach, where the shells were all cast up broken and divided and never in pairs as they are found in the

sea when alive, with two valves which form a covering the one to the other.

6. Leonardo often illustrates the so-called uniformitarian principle of using observations about current processes to infer past events. In a striking example, he notes how far a cockle can move in a day in order to understand the spatial distribution of shells in a layer of fossils:

> It does not swim, but makes a furrow in the sand, and supporting itself by means of the sides of this furrow will travel between three and four *braccia* in a day. [A *braccio,* or "arm," measured about two feet.]

7. No marine fossils have been found in regions or sediments not formerly covered by the sea.

8. When we find fossil shells broken in pieces, and heaped one upon the other, we may infer transport by waves and currents before deposition:

> But how could one find, in the shell of a large snail, fragments and bits of many other sorts of shells of different kinds unless they have been thrown into it by the waves of the sea as it lay dead upon the shore like the other light things which the sea casts up upon the land?

9. The age of a fossil shell can often be inferred from growth rings that record astronomical cycles of months or years. (Sclerochronology, or the analysis of periodicities in growth, has only become a rigorous and important subject in paleobiology during the current generation.) We can, Leonardo writes, "count on the shells of cockles and snails the numbers of months and years of their lives just as one can on the horns of bulls."

I have often quoted a favorite line from Darwin in these essays: "How can anyone not see that all observation must be for or against some view if it is to be of any service?" Leonardo's keen observations do seem to emit a wondrous whiff of modernity, but when we learn why he made his inquiries, and note how he ordered his facts, we can begin to place him into the proper context of his own world. Leonardo did not observe fossils for pure unbridled curiosity, with no aim in mind and no questions to test. He recorded all his information for a stated and definite purpose—to confute the two major interpretations of fossils current in his day. Both theories had been proposed to resolve a problem that had troubled Western natural history ever since antiquity: If fossil shells are the remains of marine organisms (and some are virtually indistinguishable from modern species), how did they get entombed in strata that now lie within mountains, several thousand feet above current sea level?

First, Leonardo disproves and ridicules the common idea that all fossils reached the mountains by transport on the high waters and violent currents of Noah's flood. Observations 3 through 6 of my list refute this theory by noting that many fossils are preserved in their position of life, undisturbed by any movement after death. One flood cannot produce a fossil record in several sequential layers (observation 3). Strata formed by violent currents could not preserve the feeding tracks of worms (observation 4). Noah's floodwaters would have disarticulated all fossil clams into separate valves (observation 5). As for the cockle, laboriously moving but six to eight feet a day in its furrow, forty days and nights of rain would scarcely provide enough time for a journey 250 miles inland (where fossil cockles now reside) from the nearest modern sea:

> With such a rate of motion it would not have traveled from the Adriatic Sea as far as Monferrato in Lombardy, a distance of 250 miles, in forty days—as he has said who kept a record of this time.

Moreover, Leonardo adds, cockle shells are too heavy to be carried at the tops of waves, while they cannot be swept up the mountains along the bottom of the waters because Leonardo believed that bottom currents always move down from higher to lower elevations, even while waves and surface currents sweep inland.

The explicit refutation of Noah's flood as a cause of fossils forms a major theme of the Leicester Codex, and occupies several full pages of text—one, for example, titled "of the Flood and of marine shells," and another, "Refutation of such as say that the shells were carried a distance of many days' journey from the sea by reason of the Deluge."

Second, Leonardo dismisses, even more contemptuously, various Neoplatonic versions of the theory that fossils are not remains of ancient organisms at all, but manifestations of some plastic force within rocks, or some emanation from the stars, capable of precisely mimicking a living creature in order to illustrate the symbolic harmony among realms of nature: animal, vegetable, and mineral. For if fossils really belong to the mineral kingdom, then their position on the tops of mountains ceases to be anomalous, as we need no longer believe that these objects ever inhabited the seas.

Leonardo made observations 7 through 9 to refute this Neoplatonic theory that fossils "grew" within their entombing rocks and do not represent the remains of organisms. If marine fossils are inorganic, why don't they "grow" in all strata, rather than only in rocks carrying abundant evidence of an oceanic origin (observation 7)? If fossils belong to the mineral kingdom, why should they so often grow in fragments and jumbles looking exactly like piles of shells on our beaches, or layers deposited by rivers in lakes and ponds (observation 8)? Most convincingly, if fossils grow from inorganic "seeds" in the rocks, how can they expand, year by year, as indicated by growth bands in their shells, without fracturing the surrounding matrix (observation 9)?

Leonardo reserved his choicest invective for what he regarded as the lingering magical content of this Neoplatonic theory of signs and signatures (although the issue remained alive—and quite lively—within Western science until the late seventeenth century. The *Mundus subterraneus* [1664] of the great Jesuit scholar Athanasius Kircher represents the last seriously cogent defense of the Neoplatonic position). Leonardo writes:

> And if you should say that these shells have been and still constantly are being created in such places as these by the nature of the locality and through the potency of the heavens in those spots, such an opinion cannot exist in brains possessed of any extensive powers of reasoning because the years of their growth are numbered upon the outer coverings of their shells [observation 9 again]; and both small and large ones may be seen, and these would not have grown without feeding or feed without movement, and here [that is, in solid rock] they would not be able to move . . . Ignoramuses maintain that nature or the heavens have created [fossils] in these places through celestial influences.

But this demonstration that Leonardo made his paleontological observations to refute the prevailing theories of his time scarcely establishes my argument that he must be evaluated as a thinker immersed in his own premodern context, and not judged for his remarkable foreshadowing of twentieth-century views—for a true spaceman would also have to refute the fallacies of his surroundings in order to introduce superior views from his time warp (just as Hank Morgan had to reject the running messenger service to favor a telephone call for summoning Sir Lancelot's bicycle corps). I must advance a further claim—one that can be particularly well documented in Leonardo's case.

Just as Leonardo made his astute observations to refute prevailing theories of fossils, he also urged his interpretations *in support of* his own favored theory of the earth. ("All observation must be *for or against* some view . . .") And the positive prod for Leonardo's paleontological observations could not have been more squarely Renaissance or late medieval, more firmly attached to his own time and concerns—and not to ours. Leonardo observed fossils as part of his quest to support a distinctive theory of the earth—a framework that would have been seriously weakened if either the Noah's flood or the Neoplatonic theory of fossils had been true. If Leonardo had not been so devoted to his "antiquated" theory of the earth, I doubt that he would ever have been inspired to make his wonderfully "modernist" observations about fossils—for the notebooks invariably present his observations as arguments to support his theory.

Leonardo was so much larger than life, even in the eyes of his contemporaries, that a potent mythology began to envelop him right from the start. Only thirty years after Leonardo's death, Giorgio Vasari published a first biography full of such touching tall tales as Leonardo's death in the arms of King Francis the First. (Francis did admire Leonardo greatly, but he and his entire court had decamped to another town on the day of Leonardo's demise. A. Richard Turner has written an entire, and fascinating, book on the history of the Leonardo legend through the ages: *Inventing Leonardo,* University of California Press, 1992.) One prominent component of the myth—that Leonardo was an unlettered man who could only work by observation and therefore gained great (if ironic) benefit from *not knowing* the false traditions of medieval Scholasticism—must be refuted if my case for his medieval impetus has merit. For how could I assert such a controlling context if Leonardo never knew or studied the prevailing traditions of book learning in his time?

As the illegitimate son of a Florentine notary, Leonardo grew up in comfortable but nonscholarly circles, and received only a limited formal education. Most important, he did not learn Latin, then the nearly universal language of intellectual communication. But Leonardo did study Latin assiduously in later life, even if he never attained more than a halting knowledge. (I love Martin Kemp's statement in his superb book *Leonardo da Vinci: The Marvelous Works of Nature and Man:* "It is rather humbling to think of Leonardo in his late thirties secretly schooling himself in the rhythmic rotes of 'amo, amas, amat . . . ,' like one of the children of the court.")

Moreover, Leonardo studied Latin because he yearned to gain full access to the scholarship of classical and medieval sources. He built a respectable library for the time—Italian translations whenever possible, but Latin when necessary. He read particularly widely and deeply in this essay's subject of paleontology and the structure of the earth. Kemp writes: "He was taking up questions which had provided considerable bones of contention in classical and medieval science. An impressive roll call of classical authorities contributed to his education in physical geography . . . There probably is no other field in which Leonardo's knowledge of classical and medieval sources was so extensive."

He read the Greek masters Aristotle and Theophrastus on geology; he owned a copy of Pliny's encyclopedic *Natural History;* he studied the views of the great Islamic scholars Avicenna and Averroes (mainly via medieval Christian sources). He listed parts of what he had read and owned on the inside front cover of his Manuscript F—Aristotle's *Meteorologia,* Archimedes on the center of gravity, "Albertuccio and Albertus *de coelo et mundo."* I found this last comment particularly sweet, as Leonardo follows medieval conventions in distinguishing his sources as "Little Al" (the Italian diminutive Albertuccio) and "Big Al." Little Al is Albert of Saxony (ca. 1316–1390), the German Scholastic philosopher and physicist. Later scholars frequently confused him with Big Al, or Albertus Magnus (ca. 1200–1280), Albert the Great, the teacher of Saint Thomas Aquinas. Both Als wrote extensively about the form and behavior of the earth, and Leonardo probably learned the views of Jean Buridan (1300–1358) by reading Albert of Saxony's discussion. Buridan's concept became the basis for the theory of the earth that Leonardo defended with his observations on fossils.

What theory of the earth, then, did Leonardo seek to support with paleontological data? Simply stated, Leonardo was vigorously promoting a common and distinctively premodern view that could not have been more central to all his thought and art: the comparison, and causal union, of the earth as a macrocosm with the human body as a microcosm. We tend to regard such comparisons today as "merely" analogical or "purely" metaphorical—more apt to promote a deluding sense of false unity than any genuine insight about common causality. By contrast, Leonardo's pre-

modern world viewed such consonances as deeply meaningful, in part by
invoking the same general theory of symbolic correspondence across scales
of size and realms of matter that Leonardo (ironically) had rejected so vigorously in denying the Neoplatonic idea that fossils might grow within
rocks as products of the mineral kingdom.

No theme recurs so incessantly, and with such central import, both in
the Leicester Codex and throughout Leonardo's writing, as the causal and
material unity of the body's microcosm and the earth's macrocosm.
Leonardo also knew the ancient pedigree of this doctrine, from classical
antiquity through medieval Scholasticism. In the A Manuscript (now in the
Institut de France), Leonardo stated that he would begin his "Treatise on
Water" (never completed or published) with a statement that he later
repeats almost verbatim in the Leicester Codex:

> Man has been called by the ancients a lesser world, and indeed the
> term is rightly applied, seeing that if man is compounded of earth,
> water, air and fire, this body of the earth is the same; and as man
> has within himself bones as a stay and framework for the flesh, so
> the world has the rocks which are the supports of the earth; as man
> has within him a pool of blood wherein the lungs as he breathes
> expand and contract, so the body of the earth has its ocean, which
> also rises and falls every six hours with the breathing of the world
> [the tides]; as from the said pool of blood proceed the veins which
> spread their branches through the human body, in just the same
> manner the ocean fills the body of the earth with an infinite num-
> ber of veins of water.

We need go no further than Leonardo's most celebrated creation—the
Mona Lisa—to recognize the analogy of macrocosm and microcosm as the
centerpiece of his thought. *La Gioconda* stands on a balcony overlooking a
complex geological background of flowing waters that complete a full
hydrological cycle just as blood moves through the human body. Martin
Kemp notes:

31

In Leonardo's Mona Lisa, note how the background represents a complex cycle of flowing water and how this flow is continued and mimicked in La Gioconda's *hair and in the folds and wrinkles of her garments (as well as, by implication, in the flow of blood within her body).*

The processes of living nature are not only mirrored by anatomical implication within the lady's body, but are more obviously echoed in the surface details of her figure and garments, which are animated by myriad motions of ripple and flow. The delicate cascades of her hair beautifully correspond to the movement of water, as Leonardo himself was delighted to observe: "Note the motion of the surface of the water which conforms to that of the hair." . . . The little rivulets of drapery falling from her gathered neckline underscore this analogy, as do the spiral folds of the veil across her left breast.

We now reach the central dilemma that makes the paleontological observations so crucial to the argument of the Leicester Codex. This notebook, as scholars have always recognized, is primarily a treatise on the nature of water in all its properties, manifestations, and uses. Why, then, does Leonardo devote so much apparently subsidiary space to the nature of fossils and the reason for their situation in mountain strata, far above present sea level? The key to this problem lies in his almost heroic struggle to overcome a central difficulty in validating his crucial analogy of the body's microcosm to the earth's macrocosm. Most scholars have missed this theme, and therefore do not grasp the union of the hydrological and paleontological passages of the Leicester Codex.

Leonardo recognizes only too well—for he has struggled with this problem for years, and through several notebooks—that his crucial analogy suffers from a potentially fatal difference between the human body and the earth. Both are constructed from the four elements of antiquity: earth, water, air, and fire. But the human body sustains itself by circulating these elements, particularly by maintaining some mechanism for permitting water (blood) to rise from the legs to the head. The analogy of microcosm and macrocosm can only work if the earth also possesses a comparable device for sustenance by cycling.

But how can such a notion be defended for the planet, especially in the face of an apparently disabling problem: earth and water are heavy elements; their natural motion must produce a downward flow (leading ideally to a planet of four concentric layers with earth at the center, water above, air atop water, and fire at the periphery). If earth and water must move down, these heavy elements will eventually stabilize as two concentric spheres at the center of the planet—and the macrocosm will therefore possess no device for sustenance by circulation. Leonardo knows that he must therefore find a mechanism to make both earth and water *move up* (against their natural tendency), as well as down, on our planet. This pressing need, so difficult to validate, sets the central struggle that engages Leonardo throughout the Leicester Codex.

Ironically, I wish to argue, he never did solve his problem for the main subject of the codex—water. That is, he tried again and again, but never found a satisfactory mechanism to guarantee the upward motion, hence the cycling, of water. However—and we now come to the crucial point that has usually been missed—Leonardo did succeed (by his standards) in the quest to find a mechanism for upward movement of the other heavy element: earth. Fossils on mountains provide the observational proof that earth can rise, both generally and often, for marine shells once inhabited the sea but now reside in the high mountains. The paleontological observations form a centerpiece in the Leicester Codex not, as has usually been argued, because fossils once lived in water and the codex treats water in all major aspects (an awfully lame reason for devoting so much space to paleontology), but rather because fossils record Leonardo's great success (in contrast to his failure for the central subject of water)—his key evidence for a general mechanism to drive the upward motion of earth, and therefore, his proof for a self-sustaining planet that may legitimately be compared with the human body.

Leonardo knew only too well that he faced a serious problem with the motion of water through the earth, and he virtually obsesses over the issue in notebook after notebook, repeating the conundrum in almost unchanging words, and proposing various solutions, only to abandon them later as untenable. Water, by itself and following its "natural course" (Leonardo's words), can only flow down. But within the earth, water must also move up along internal channels (comparable with blood vessels of the human body) to emerge as springs in the high mountains (and thence, back on track, to flow as rivers to the sea). An earthly force must therefore make water rise through the land against its natural inclination to flow down. The combined action of these two forces will cause water to circulate—and thereby act like the blood in our bodies to sustain a living system.

> So does the water which is moved from the deep sea up to the summits of the mountains, and through the burst veins [mountain springs] it falls down again to the shallows of the sea, and so rises again to the height where it burst through, and then returns in the same descent. Thus proceeding alternately upwards and downwards at times it obeys its own desire [to move down] at times that of the body in which it is pent [to move up]. (From the Arundal Codex in the British Museum.)

Leonardo could not have been more explicit in admitting that water can move upward only by running against its natural course and that, if any mechanism can be found at all for this anomalous motion, the analogy between microcosm and macrocosm offers the only reasonable hope:

> Clearly it would seem that the whole surface of the oceans, when not affected by the tempest, is equally distant from the center of the earth, and that the tops of the mountains are as much farther removed from this center as they rise above the surface of the sea.[3] Unless therefore the body of the earth resembled that of man it would not be possible that the water of the sea, being so much lower than the mountains,

_____

3. Here Leonardo properly rejects a popular explanation supported by some of his contemporaries—that the earth, in cross section, is an elongated ellipse rather than a sphere, and that water at the end of the long axis of the ellipse will stand higher (farther from the earth's center) than mountains in the short axis of the ellipse. Water could then flow "down" from oceans to mountains. But Leonardo recognized that the surface of the ocean will always lie below the summits of mountains on other quadrants of the earth's surface.

should have power in its nature to rise to the summit of the mountains. We must therefore believe that the same cause that keeps the blood at the top of a man's head keeps water at the summit of the mountains. (From the A Manuscript in the Institut de France.)

But to state a need is not equivalent to finding a mechanism. Throughout the Leicester Codex, Leonardo struggles to discover a physically workable way to raise water within the earth. He tries and rejects several explanations, as Martin Kemp documents in an article entitled "The Body of the Earth." Perhaps, Leonardo first argues, the heat of the sun draws water up through the veins (internal streams) that run through mountains. (Leonardo, in his strongest image of a living earth, had written in the Leicester Codex: "The body of the earth, like the bodies of animals, is interwoven with a network of veins, which are all joined together, and formed for the nutrition and vivification of the earth and of its creatures.") But he then realizes that this explanation cannot work for two reasons—first, because, on the highest mountaintops closest to the heating sun, water remains cold, and even icy; and, second, because this mechanism should operate best in summer during maximal solar heat, but mountain rivers often run with lowest waters at this time.

In a second try, Leonardo turns to the earth's internal heat and a process of distillation: perhaps the interior fires boil water in internal caverns, and this water rises as vapor through mountain interiors, where it reverts to liquid form and bursts through as a high spring. But this proposal won't work either because such extensive distillation would require that the roofs of internal caverns be wet with the rising steam—but they are often bone-dry. Leonardo then made a feebler third attempt: perhaps, by analogy to a sponge, mountains somehow suck up water to a point of saturation and subsequent oozing from the top. But Leonardo realizes that he cannot cash out this analogy in mechanical terms:

> If you should say that the earth's action is like that of a sponge which, when part of it is placed in water, sucks up the water so that

it passes up to the top of the sponge, the answer is that even if the water itself rises to the top of the sponge, it cannot then pour away any part of itself down from this top, unless it is squeezed by something else, whereas with the summits of the mountains one sees it is just the opposite, for there the water always flows away of its own accord without being squeezed by anything.

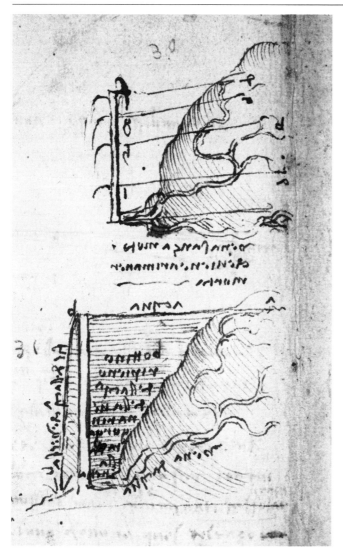

Leonardo's sketches of internal rivers within the body of the earth. These rivers allow water to rise to the tops of the mountains.

(One may wonder, of course, why Leonardo doesn't invoke an explanation recognized in his day, and now known as "obviously" correct—that water "moves up" as evaporated vapor, later to fall as rain on mountaintops. In fact, Leonardo reluctantly acknowledged this resolution in notebooks written later than the Leicester Codex. But, when we get "inside" Leonardo's head and his own explanatory world, we can easily see why he would shun a resolution that now seems so obvious to us. Leonardo wanted to prove that water in a living earth moves like blood in a living body—and this analogy required that water flow both up and down *within* earthly channels that could be likened to blood vessels. Blood does not evaporate and fall as rain in our heads!)

But if Leonardo, to his great disappointment, never solved the problem of rising waters, he did (to his satisfaction) crack the equally knotty problem of a general mechanism for the elevation of earth—a combination of his views on gravity and his concept of erosion. (I struggled with Leonardo's complex mix of ideas for many days—a mélange of scholastic theories of gravity and the earth, mainly vouchsafed to Leonardo by Jean Buridan through the books of Albert of Saxony, and of Leonardo's conjectures on composition of the earth's interior combined with observations on our planet's surface—but I am now confident that I grasp the argument and can present a crisp epitome.)

Our planet has a geometric center, called by Leonardo the "center of the world" or sometimes the "center of the universe"—for Leonardo predated Copernicus and accepted the Ptolemaic system of a central earth and a revolving sun. The realm of liquid water must arrange itself as a perfect sphere about this center, with the surface of the ocean equidistant at all points from the center of the world. If solid earth were homogeneous and equally distributed, this element would also form a smooth sphere with a surface equidistant at all points from the center of the world. (By the way, and contrary to popular mythology, all scholars recognized—and had known since antiquity—that the earth was spherical and not flat.)

But the heavy earth is far from homogeneous. The interior of our planet is a complexly marbled mass composed of solid earth, liquid water running through veins in the rocks, and even air, where water has hollowed out caverns in the rocks. Therefore, as a result of this unequal distribution of earth, one hemisphere must always be heavier than the other.

Now the planet also has a center of mass (called by Leonardo, in a terminology that we would not use today, a "center of gravity"). On a homogeneous planet, this "center of gravity" will coincide with the geometric center of the world. But on our actual planet, with one hemisphere heavier than the other, the "center of gravity" will lie *below* the geometric center and within the heavier hemisphere. The planet, as a living body seeking balance, must strive to bring the center of gravity closer to the geometric center. The earth pursues this goal in a manner known from time immemorial to all riders on seesaws (the Leicester Codex contains a picture of such a seesaw, albeit for a different purpose). To balance a seesaw, the heavier person must move toward the fulcrum at the center, while the lighter person must move farther away. In exactly the same manner, the solid masses of the heavier hemisphere must sink toward the center of the world, while the rocks of the lighter hemisphere must rise. The emergence of mountains from the seas, and the consequent placement of marine fossils on high hills, records this rising of land in the earth's lighter hemisphere.

Leonardo's sketch of men on a seesaw from the Leicester Codex, illustrating his understanding of the principles of weight, distance, and balance.

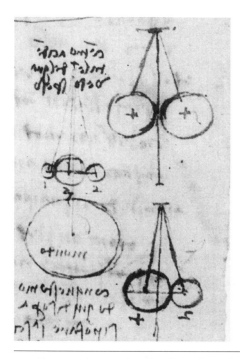

Leonardo uses these two drawings (on the right-hand side of his sketch) to show that when two objects (representing the hemispheres of the earth in his general argument) are the same weight, the vertical line (representing the center of gravity) falls directly between them. When the two objects are of different weights, the vertical line lies within the heavier object, which then forces the lighter object to rise (as the lighter hemisphere of the earth must rise).

Leonardo succinctly describes the general process in Manuscript F (in the Institut de France):

> Because the center of the natural gravity of the earth ought to be in the center of the world, the earth is always growing lighter in some part, and the part that becomes lighter pushes upwards, and submerges as much of the opposite part as is necessary for it to join the center of its aforesaid gravity to the center of the world; and the sphere of the water keeps its surface steadily equidistant from the center of the world.

Leonardo must then find a general mechanism for ensuring planetary balance by lightening one hemisphere, while making the other heavier— and he succeeds with two principles, both based on erosion by water: one mode operating in the earth's interior, the other at the earth's surface. In the interior, internal veins of water carve out caverns, which eventually become unstable. Their tops finally collapse, and enormous blocks of rock fall all the way to the center of the world. There, the blocks distribute themselves about the center with approximately equal volume in each hemisphere—thus adding weight to one hemisphere and subtracting from the other (for the entire block had previously resided in one hemisphere alone). Leonardo includes a striking illustration of this process in the Leicester Codex—although scholars have failed to recognize the meaning of this figure—showing a fallen block as a large arch neatly draped about the center of the world (see accompanying figure). In describing this internal mechanism in the Leicester Codex, Leonardo explicitly cites the rising of fossiliferous strata as a consequence:

> The fact of the summits of the mountains projecting so far above the watery sphere may be due to the fact that a very large space of the earth which was filled with water, that is the immense cavern, must have fallen in a considerable distance from its vault towards the center of the world, finding itself pierced by the course of the springs,

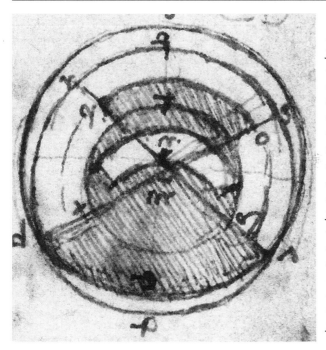

Leonardo's mechanism for the rising of mountains: the upper hemisphere becomes lighter when a large mass of earth falls to the center of the planet. The large mass is the light crescent-shaped wedge surrounding the earth's center. This wedge has fallen from a corresponding space (now darkened but of the same shape) in the upper hemisphere (the third concentric layer down from the surface).

> which continually wear away the spot through which they pass . . .
> Now this great mass has the power of falling . . . It balances itself
> with equal opposing weights round the center of the world, and
> lightens the earth from which it is divided; and it [the lightened
> earth] removed itself immediately from the center of the world and
> rose to the height, for so one sees the layers of the rocks [with their
> fossils], formed by the changes which the water has undergone, at
> the summits of the high mountains.

The exterior method of lightening by erosion can enhance this process
once the mountains rise. Rivers will now erode the sides of the mountains
and carry the resulting sediment away to the oceans. Some of this sediment
will flow to the opposite hemisphere, thus further increasing the imbalance
of weight, and causing the mountains to rise still higher as a consequence.

> And now these beds are of so great a height that they have become
> hills or lofty mountains, and the rivers which wear away the sides of
> these mountains lay bare the strata of the shells, and so the light sur-
> face of the earth is continually raised, and the antipodes [the oppo-
> site side of the earth] draw nearer to the center of the earth, and the
> ancient beds of the sea become chains of mountains.

Thus, and finally, we grasp the central importance of Leonardo's paleontological observations in the Leicester Codex. He featured fossils in
order to validate the cherished centerpiece of his premodern worldview—
the venerable argument, urged throughout classical and medieval times,
for interpreting the earth as a living, self-sustaining "organism," a macrocosm working by the same principles and mechanisms as the microcosm of
the human body. Leonardo required, above all, a general device to make
the heavy elements, earth and water, move upward against their natural
inclination—so that the earth could sustain itself, like a living body, by
constantly cycling all its elements, rather than reaching inert stability with
heavy elements in permanent layers below lighter elements.

Leonardo could not find such a mechanism for the chief subject of the Leicester Codex: water—and this failure caused him great frustration. But he succeeded for the even heavier element of earth. He extended a mechanism proposed by Scholastic philosophers for causing the lighter hemisphere of an inhomogeneous planet to rise. He proposed both internal and external erosion by water as processes that could lighten a hemisphere—but he needed observational evidence that land did, in fact, rise. His crowning jewel of confirmation lay in a well-known phenomenon that had provoked intense debate ever since the days of classical Greek science—fossils of marine organisms in strata on high mountains.

Leonardo also needed to assert that the elevation of strata with fossils must represent a general and repeatable feature of the earth's behavior, not an odd or anomalous event. Thus he had to refute the two explanations for fossils most common in his time—for Noah's flood could only be viewed as a strange and singular phenomenon, and if all fossils derive from this event, then paleontology illustrates no general mechanism for the rising of land. And if fossils grow as objects of the mineral kingdom within rocks, then the mountains may always have stood high, and we can derive no evidence for any uplift at all. Thus Leonardo made his superb observations on fossils in order to validate his lovely, but ever so antiquated, view of a causally meaningful and precise unity between the human body as a microcosm and the earth as a macrocosm. Leonardo, the truly brilliant observer, was no spaceman, but a citizen of his own instructive and fascinating time.

I like to contemplate Leonardo, this complex man of peace, of gentleness, of art, of scholarship; this military engineer who designed (but generally did not build) ingenious instruments of war, but who would not reveal his ideas for a submarine, as he stated in the Leicester Codex:

> This I do not publish or divulge on account of the evil nature of men who would practice assassinations at the bottom of the seas, by breaking the ships in their lowest parts and sinking them together with the crews who are in them.

And I like to compare his views on the mechanism for raising mountains from the sea (and exposing fossils for collectors) with our most celebrated literary image on the same subject—Isaiah's prophecy that "every valley shall be exalted." I also recall the peace that shall reign on Isaiah's mountain (festooned, no doubt, with fossils), where a scholar might study the raising of earth to his heart's content, and not need to provide his warlike patron with plans for the raising of sieges or the razing of enemy cities. Isaiah's summit, where "the wolf also shell dwell with the lamb, and the leopard shall lie down with the kid . . . They shall not hurt nor destroy in all my holy mountain."

2

THE *GREAT WESTERN*
AND THE FIGHTING
TEMERAIRE

SCIENCE PROGRESSES; ART CHANGES. SCIENTISTS ARE INTERCHANGEABLE and anonymous before their universal achievements; artists are idiosyncratic and necessary creators of their unique masterpieces. If Copernicus and Galileo had never lived, the earth would still revolve around the sun, and earthlings would have learned this natural truth in due time. If Michelangelo had never lived, the Sistine Chapel might still have a painted vault, but the history of art would be different and humanity would be a good deal poorer. This "standard" account of the differences between art and science belongs to our distressing but prevalent genre of grossly oversimplified dichotomies— stark contrasts that both enlighten in their boldness and distort in their formulaic divisions of complexly intertwined entities into two strictly separated piles—"and never the twain shall meet, / Till Earth and Sky stand presently at God's great Judgment Seat."

The supposed inexorability of technological progress, under this distorting dichotomy, leads to the myth of science as virtually disembodied—a machine endowed with its own momentum, and therefore striding forward almost independently of any human driver. Scientists, under this model, become anonymous and virtually invisible. A few names survive as icons and heroes—Edison and Bell as doers, Darwin and Einstein as thinkers. But, if we accept the premise that technological innovation (in manufacturing, warfare, transportation, and communication) has powered social change far beyond all other consequences of human emotion and ingenuity, how can we resolve the paradox that the people most responsible for propelling human history remain so invisible? Who can name anyone connected with the invention of the crossbow, the zipper, the typewriter, the Xerox machine, or the computer?

Artists, politicians, and soldiers win plaudits and notoriety, though so many impose themselves only lightly and transiently upon the motors of social change. Scientists, engineers, and technologists forge history and gain oblivion as a reward—in large part as a consequence of the false belief that individuality has little relevance when a progressive chain of discoveries proceeds in logical and inexorable order. Let me illustrate our different treatment of scientists versus statesmen and artists with two pairings.

Colonel Calverly, head of a company of dragoon guards in Gilbert and Sullivan's *Patience,* introduces his troops by giving the audience a formula for their construction:

> *If you want a receipt for that popular mystery,*
> *Known to the world as a heavy dragoon,*
> *Take all the remarkable people in history,*
> *Rattle them off to a popular tune . . .*

The Colonel then rips off (at patter-song speed) two hilarious doggerel verses, listing thirty-eight historical figures, including a few fictional and general characters. Only one is a scientist. (The notoriously sexist Gilbert listed three times as many women—Queen Anne, the generic and demeaning "Odalisque on a divan," and Madame Tussaud, founder of the great London wax museum.) The scientist appears in the first quatrain:

> *The pluck of Lord Nelson on board of the* Victory—
> *Genius of Bismarck devising a plan—*
> *The humor of Fielding (which sounds contradictory)—*
> *Coolness of Paget about to trepan.*

Most of us will have no trouble with the first three—Admiral Horatio Nelson dying at the battle of Trafalgar, the great German statesman, and the author of *Tom Jones*. But scientists gain little recognition in their own times and quickly fade from later memory. So who is Mr. Paget, about to open his patient's skull? Sir James Paget, surgeon to the queen and a founder of the science of pathology, may have been a household name to his Victorian contemporaries, but few of us know him today (and I couldn't have made the identification without my trusty encyclopedia). So scientists and engineers create history, but Gilbert chooses only one to participate in the construction of English fiber, and even this man has since sunk to oblivion in the general culture of educated people.

For the second pairing, let us return to Admiral Nelson and the story of Trafalgar. On October 21, 1805, Nelson's fleet of twenty-seven ships met and destroyed a combined French and Spanish force of thirty-three vessels off Cape Trafalgar, near the Strait of Gibraltar. Nelson's forces captured twenty ships and put 14,000 of the enemy out of commission (about half killed or wounded, and half captured), while suffering only 1,500 casualties and losing no ships. This victory ended Napoleon's threat to invade England and established a supremacy of British naval power that would endure for more than a century.

Nelson, "on board of the *Victory,*" engaged his flagship with the French *Redoutable.* The opposing ship fired at such close range that a French sniper, shooting from the mizzentop of the *Redoutable,* easily picked off Nelson from a distance of only fifteen yards. Nelson died of this wound a few hours later, but with secure knowledge of his triumph.

Nelson's ship, and much of the battle, was saved by the second man-of-war on the line, the *Temeraire*. This vessel rescued the *Victory* by firing a port broadside into the *Redoutable* and disabling the French ship. (The mainmast of the *Redoutable* fell right across the *Temeraire;* the French ship then surrendered, and the *Temeraire*'s crew boarded her and lashed the defeated vessel to her port side.) Another French ship, the *Fougueux,* then attacked the *Temeraire,* but the British man-of-war fired her starboard broadside, to equally good effect, and secured her second prize, lashed this time to her starboard side. The *Temeraire,* now disabled herself, but with her two prizes lashed to her sides, had to be towed into port by a frigate.

Enter J. M. W. Turner (1775–1851), Britain's greatest nineteenth-century artist and the first subject of my second pairing. Early in his career, in 1806, Turner painted a conventionally heroic scene of the conflict: *The Battle of Trafalgar, as Seen from the Mizen Starboard Shrouds of the* Victory. We see Nelson, surrounded by his officers and dying on deck. The *Temeraire* stands in the background, firing away at the *Fougueux.*

Late in his career, in 1839, Turner returned to the ships of Trafalgar and depicted a very different scene, magnificent in philosophical and emotional meaning, and one of the world's most popular paintings ever since: *The Fighting* Temeraire, *Tugged to Her Last Berth to Be Broken Up, 1838.* The large men-of-war, with their three major tiers of guns, were beautiful, terrible (in the old sense of inspiring terror), and awesome fighting machines. The *Temeraire,* constructed of oak, was built at Chatham and launched in 1798. The ship carried a crew of 750, far more than needed to sail the ship (with a gundeck 185 feet in length), but required to operate the ninety-eight guns—for each gun employed several men in elaborate procedures of loading, aiming, firing, and controlling the recoil. But these "hearts of oak" (the favored patriotic name for the great men-of-war) fell victim to their own success. Their supremacy removed the threat of future war, while advancing technologies of steam and iron soon outpaced their wood and sails. These ships never fought again after the Napoleonic wars, and most were reduced to various workaday and unsentimental duties in or near port. The

Temeraire, for example, was decommissioned in 1812 and then served as a floating prison and a victualing station.

Eventually, as timbers rotted and obsolescence advanced, these great vessels were stripped and sold to ship breakers to be dismantled for timber, plank by plank. John Beatson, a ship breaker at the yards of Rotherhithe, bought the *Temeraire* at auction for 5,530 pounds. Two steam tugs towed the hulk of the *Temeraire* fifty-five miles from Sheerness to Rotherhithe in September 1838.

J. M. W. Turner's The Fighting Temeraire Tugged to Her Last Berth to Be Broken Up, *1838.*

Turner's painting presents a wrenchingly dramatic view, quite inaccurate in an entirely studied way, of the *Temeraire*'s last sad trip. The great man-of-war, ghostly white, still bears its three masts proudly, with light rigging in place, and sails furled on the yards. The small steam tug, painted dark red to black, stands in front, smoke belching from its tall stack to obscure part of the *Temeraire*'s mast behind. One of Turner's most brilliant sunsets—with clear metaphorical meaning—occupies the right half of the painting. The most majestic and heart-stopping product of the old order sails passively to her death, towed by a relatively diminutive object of the new technology. John Ruskin wrote: "Of all pictures not visibly involving human pain, this is the most pathetic that ever was painted."

Turner clearly set his scene for romance and meaning, not for accuracy. Ships sold for timber were always demasted, so the *Temeraire* sailed to her doom as a hulk without masts, sails, or rigging of any sort—a most uninspiring, if truthful, image. Moreover, Rotherhithe lies due west of Sheerness, so the sun never could have set *behind* the *Temeraire*!

A simplistic and evidently false interpretation has often been presented for Turner's painting—one that, if true, would establish bitter hostility between art and science, thus subverting the aim of this essay: to argue that the two fields, while legitimately separate in some crucial ways, remain bound in ties of potentially friendly and reinforcing interaction. In this adversarial interpretation, recalling Blake's contrast of "dark Satanic mills" with "England's green and pleasant land," the little steam tug is a malicious enemy—a symbol of technology's power to debase and destroy all that previous art had created in nobility. In a famous, if misguided, assessment, William Makepeace Thackeray (one of the thirty-eight in Gilbert's recipe for a heavy dragoon), wrote in 1839, when Turner first displayed his painting:

> "The Fighting Temeraire"—as grand a picture as ever figured on the walls of any academy, or came from the easel of any painter. The old Temeraire is dragged to her last home by a little, spiteful, diabolical steamer . . . The little demon of a steamer is belching out a volume . . . of foul, lurid, red-hot malignant smoke, paddling furiously, and lashing up the water around it; while behind it . . . slow, sad and majestic, follows the brave old ship, with death, as it were, written on her.

This reading makes little sense because Turner, like so many artists of the nineteenth century, was captivated by new technologies, and purposefully sought to include them in his paintings. In fact, Turner had a special fascination for steam, and he clearly delighted in mixing the dark smoke of the new technology with nature's lighter daytime colors.

In *Turner: The Fighting* Temeraire, art historian Judy Egerton documents Turner's numerous, and clearly loving, paintings of steam vessels—starting with a paddle-steamer shown prominently in a painting of Dover Castle in 1822 (passenger steamboats only started to operate between Calais and Dover in 1821), and culminating in a long series of paintings and drawings featuring steamboats on the Seine, and done during the 1830s. A perceptive commentator, writing anonymously in the *Quarterly Review* in 1836, praised Turner for creating "a new object of admiration—a new instance of the beautiful—the upright and indomitable march of the self-impelling steamboat." He then specifically lauded "the admirable manner in which Turner, the most ideal of our landscape painters, has introduced the steamboat in some views taken from the Seine."

This reviewer then credits Turner for his fruitful and reinforcing union of nature and technology:

> The tall black chimney, the black hull, and the long wreath of smoke left lying on the air, present, on this river, an image of life, and of majestic life, which appears only to have assumed its rightful position when seen amongst the simple and grand productions of nature.

The steam tug in *The Fighting* Temeraire is not spiteful or demonic. She does not mock her passive burden on the way to destruction. She is a little workaday boat doing her appointed job. If Turner's painting implies any villain, we must surely look to the bureaucrats of the British Admiralty who let the great men-of-war decay, and then sold them for scrap.

Which brings me to Isambard Kingdom Brunel, the engineer who goes with Turner in my second pairing. How many of you know his name? How many even recognized the words as identifying a person, rather than a tiny principality somehow never noticed in our atlas or stamp album? Yet one can make a good argument—certainly in symbolic terms for the enterprise he represented, if not in actuality for his personal influence—that Isambard Kingdom Brunel was the most important figure in the entire nineteenth-century history of Britain.

Brunel was the greatest practical builder and engineer in British industrial history—and industry powered the Victorian world, often setting the course of politics as firmly as the routes of transportation. Brunel (1806–1859) built bridges, docks, and tunnels. He constructed a floating armored barge, and designed the large guns as well, for the attack on Kronstadt during the Crimean War. He built a complete prefabricated hospital, shipped in sections to the Crimea in 1855.

But Brunel achieved his greatest impact in the world of steam, both on land and at sea—and now we begin to grasp the tie to Turner. He constructed more than one thousand miles of railroad in Great Britain and Ireland. He also built two railways in Italy and served as adviser for other lines in Australia and India. In the culmination of his career, Brunel constructed the three greatest steam vessels of his age, each the world's largest at launching. His first, the *Great Western,* establishes the symbolic connection with Turner and *The Fighting* Temeraire. The *Great Western,* a wooden paddle-wheel vessel 236 feet in length and weighing 1,340 tons, was the first steamship to provide regular transatlantic service. She began her crossings in 1838, the year of the *Temeraire*'s last tow and demise. In fact, on August 17, 1838, the day after the sale of the *Temeraire,* the *Great Western* arrived in New York and the *Shipping and Mercantile Gazette* declared that "the whole of the mercantile world . . . will from this moment adopt the new conveyance." The little tug in Turner's painting did not doom or threaten the great sailing ships. Brunel's massive steam vessels signaled the inevitable end of sail as a principal and practical method of oceanic transport.

Brunel went on, building bigger and better steamships. He launched the *Great Britain* in 1844, an iron-hulled ship 322 feet long, and the first large steam vessel powered by a screw propeller rather than side paddles. Finally, in 1859, Brunel launched the *Great Eastern,* with a double iron hull and propulsion by both screws and paddles. The *Great Eastern* remained the world's largest steamship for forty years. She never worked well as a passenger vessel, but garnered her greatest fame in laying the first successful transatlantic cable. Brunel, unfortunately, did not live to see the *Great Eastern* depart on her first transoceanic voyage. He suffered a serious stroke on board the ship, and died just a few days before the voyage.

Turner and Brunel are bound by tighter connections than the fortuitous link of the *Temeraire*'s demise with the inauguration of regular transoceanic service by the *Great Western* in the same year of 1838. Turner also loved steam in its major manifestation on land—railroads. In 1844, his seventieth year, Turner painted a canvas that many critics regard as his last great work: *Rain, Steam, and Speed—The Great Western Railway.* Brunel built this two-hundred-mile line between London and Birmingham between 1834 and 1838 (and then used the same name for his first great steamship). Turner's painting shows a train, running on Brunel's wide seven-foot gauge, as the engine passes over the Maidenhead Railway Bridge, another famous construction, featuring the world's flattest brick arch, as designed and built by Isambard Kingdom Brunel. The trains could achieve speeds in excess of fifty miles per hour, but Turner has painted a hare running in front of the engine—and, though one can't be sure, the hare seems poised to outrun the train, not to be crushed under "the ringing grooves of change," to cite Tennyson's famous metaphor about progress, inspired by the poet's first view of a railroad.

We revere Turner, and rightly so. But why has the name of Isambard Kingdom Brunel, as inspired in engineering as Turner in painting, as influential in nineteenth-century history as any person in the arts, slipped so far from public memory? I do not know the full answer to this conundrum, but the myth of inexorability in discovery, ironically fostered by science as a source of putative prestige, has surely contributed by depicting scientists as interchangeable cogs in the wheel of technological progress—as people whose idiosyncracy and individual genius must be viewed as irrelevant to an inevitable sequence of advances.

Art and science are different enterprises, but the boundaries between them remain far more fluid and interdigitating, and the interactions far richer and more varied, than the usual stereotypes proclaim. As a reminder of both overlaps and differences, I recently read the first issue of *Scientific American*—for August 28, 1845, and republished by the magazine to celebrate its 150th anniversary.

Scientific American was founded by Rufus Porter, a true American original in eccentric genius and entrepreneurial skill. Porter had spent most of his time as an itinerant mural painter, responsible for hundreds of charming and primitively painted landscape scenes on the interior walls of houses throughout New England. Yet he chose to start a journal devoted primarily to the practical side of science in engineering and manufacturing. In fact, the initiating issue features, as the main article, a story about the first landing in New York of "the greatest maratime [*sic*] curiosity ever seen in our harbour"—none other than Brunel's second ship, the *Great Britain*. "This mammoth of the ocean," Porter writes, "has created much excitement here as well as in Europe . . . During the first few days since her arrival at New York, she has been visited by about 12,000 people, who have paid 25 cents for the gratification."

If an artist could initiate a leading journal in science, if Turner could greatly enhance his painted sunsets by using a new pigment, iodine scarlet, just invented by Humphrey Davy of the Royal Institution, a leading scientific laboratory founded by Count Rumford in 1799, then why do we so consistently stress the differences and underplay the similarities between these two greatest expressions of human genius? Why do we pay primary attention to the artist's individuality, while constantly emphasizing the disembodied logic of science? Aren't these differences of focus mostly a matter of choice and convention, not only of evident necessity? The individuality of scientists bears respect and holds importance as well. I do accept that we would now know about evolution even if Darwin had never been born. But the discovery would then have been made by other people, perhaps in different lands, and surely with dissimilar interests and concerns—and these potential variations in style may be no less profound or portentous than the disparity between such artistic contemporaries as Verdi and Wagner.

I do not deny that the accumulative character of scientific change—the best justification for a notion of progress in human history—establishes the major difference between art and science. I found a poignant reminder within a small item in the first issue of *Scientific American.* An advertisement for daguerreotypes on the last page includes the following come-on: "Likenesses of deceased persons taken in any part of the city and vicinity." I then remembered a book published a few years ago on daguerreotypes of dead children—often the only likeness that parents would retain of a lost son or daughter. (Daguerreotypes required long exposures, and young children could rarely be enticed to sit still for the requisite time—but the dead do not move, and daguerreotypists therefore maintained a thriving business, however ghoulish by modern standards, in images of the deceased, particularly of children.)

No example of scientific progress can be less subject to denial or more emotionally immediate than our ever-increasing ability to prevent the death of young people. Even the most wealthy and privileged parents of Turner and Brunel's time expected to lose a high percentage of their children. As Brunel built his railways and Turner painted, Darwin's geology teacher, Adam Sedgwick, wrote to a friend about the achievements of his young protégé, then sailing around the world on the *Beagle,* and therefore in constant medical danger, far from treatment in lands with unknown diseases: "[He] is doing admirable work in South America, and has already sent home a collection above all price . . . There was some risk of his turning out an idle man, but his character will now be fixed, and if God spares his life he will have a great name among the naturalists of Europe." A concerned mentor would not need to fret so intensely today—a blessing from science to all of us.

I previously quoted the beginning of Colonel Calverly's recipe for a heavy dragoon, and will now close with the end:

Beadle of Burlington—Richardson's show—
Mr. Micawber and Madame Tussaud!

We know Mr. Micawber from *David Copperfield,* and Mme. Tussaud for her wax statues. "Richardson's show" puzzled me until I found the following entry in my 1897 edition of *Chambers's Biographical Dictionary:* John Richardson, 1767–1837, "the 'penny showman' from Marlow work house who rose to become a well-to-do travelling manager." But who, or what, is the Beadle of Burlington?

I fell in love with Gilbert and Sullivan at age twelve, and have therefore been wondering about that Beadle for forty years (not always actively, to be sure!). Then, six months ago and to my utter delight, I ran right into the Beadle of Burlington when no subject could have been farther from my mind. I was walking down an early-nineteenth-century shopping arcade, just off Piccadilly in London, on my way to a meeting at the Royal Institution, where Humphrey Davy had invented Turner's new pigment. Lord George Cavendish founded the Burlington Arcade in 1819 "for the gratification of the public" and "to give employment to industrious females" in the shops. Lord George established firm rules of conduct for people moving through the arcade—"no whistling, singing, hurrying, humming, or making merry." Such decent standards have to be enforced—and so they have been, ever since 1819, by a two-man private security force, the Beadles of Burlington. Traditions must be maintained, of course, and the Beadles still wear their ancient garb of top hat, gloves, and coat with tails.

I looked at one of the Beadles in all his antiquated splendor, and I saw that he held both hands clasped behind his back. So I moseyed around to his other side (no hurrying) to find out what he might be holding—and I noted a cellular phone in his gloved hands. Technology and tradition. The old and elegant; the new and functional. The Fighting *Temeraire* and the steam tug. Art and science. The prophet Amos said, "Can two walk together, except they be agreed?"

3

SEEING EYE TO EYE, THROUGH A GLASS CLEARLY

WE LAUGH AT THE STUFFINESS OF VICTORIAN PRONOUNCEMENTS, AS TYPI-fied by the quintessential quotation from the woman who gave her name to the age—the Queen's reaction to an imitation of herself by her groom-in-waiting (as stated in the regal first person plural): "We are not amused." Yet we (that is, all of us poor slobs today, not her majesty alone) must also admire the unquestioned confidence, in matters both moral and material, of our Victorian forebears, especially from the ambivalent perspective of our own unsure and fragmented modernity.

In a popular book of the mid-1850s, Shirley Hibberd (an androgynous name, but male in this case, as for most publically eminent Victorians), praised the acme that his age had achieved, not only in larger affairs of state, but in the domestic tranquillity of homes as well:

> Our rooms sparkle with the products of art, and our gardens with the curiosities of nature. Our conversation shapes itself to ennobling

themes, and our pleasures take a tone from our improving moral sentiments, and acquire a poetic grace that reflects again upon both head and heart.

Hibberd argues for an intimate tie between happy homes and triumphant governments, for "our domestic life is a guarantee of our national greatness." But how shall such purity and edification be achieved on the home front? Hibberd appeals to the concept of taste:

> A Home of Taste is a tasteful home, wherein everything is a reflection of refined thoughts and chaste desires . . . In such a home Beauty presides over the education of the sentiments, and while the intellect is ripened by the many means which exist for the acquisition of knowledge, the moral nature is refined by those silent appeals of Nature and of Art, which are the foundations of Taste.

Since Hibberd was a nature writer by profession, and since I am quoting from his most famous work, titled *Rustic Adornments*, readers will not be surprised by his primary prescription for domestic improvement: the enhancement of good taste by cultivated displays of living things. "The Rustic Adornments of the household," Hibberd asserts, "embrace the highest of its attractions apart from the love which lights the walls within." Hibberd could not have been more sanguine about the beneficial moral effects of an interest in natural objects: "It would be an anomaly to find a student of nature addicted to the vices that cast so many dark shadows on our social life; nor do I remember among the sad annals of criminal history, one instance of a naturalist who became a criminal, or of a single gardener who has been hanged." (So much for the Bird Man of Alcatraz!) Moreover, an interest in nature defines both our tranquillity and our prosperity—no strife or ignobility please, we're British!

> It is because we are truly a domestic people, dearly attached to our land of green pastures, and shrubby hedgerows, and grey old woods, that we remain calm amid the strife that besets the states

around us, proud of our ancient liberties, our progressing intelligence, and our ever-expanding material resources.

But nature has always been "out there" for our edification on her turf. The greatest advance of his age, Hibberd argued, lay in the invention of devices—rustic adornments—that allowed home-dwellers, even of modest means in highly urban settings, to cultivate nature *within* domestic walls. Hibberd's book contains successive chapters on all forms of indoor natural display, from fern cases to aviaries to floral arrangements. But he devoted his opening chapter to the great craze that defined his decade of the 1850s— the establishment of marine aquariums in almost any home coveting a cachet of modernity. "I commence," Hibberd writes, "with the Aquarium, which, for its novelty, its scientific attractions, and its charming elegance, deservedly takes the first place among the Adornments of the House."

Aquariums seem so humble in concept and so common in occurrence— a staple of your dentist's office or your kid's bedroom—that we can hardly imagine an explicit beginning, or a concept of original excitement and novelty. In fact, the aquarium had a complexly interesting and particular birth during the mid-nineteenth century, and then enjoyed (or endured) one of Victorian Britain's most intense crazes of popularity during a definite interval in the 1850s. I do not, of course, claim that this invention marks the first domestic display of aquatic organisms. The owner of any respectable Roman villa could look down upon the animals in his fishpond. Similarly, the simple bowl had allowed, also since classical times, the contemplation of a fish or two in the more direct, edge-on, eye-to-eye orientation (through glass, or some other transparent medium that did not always come easily or cheaply before the last few generations).

But these precursors are not aquariums in the technical sense, for they lack the defining feature: a *stable community* of aquatic organisms that can be viewed, not from above through the opacity of flowing waters with surface ripples, but eye-to-eye and from the side through transparent glass and clear water.

A fishbowl presents a temporary display, not a stable community. The water quickly goes foul and must be changed frequently (engendering the amusing and frustrating problem, so well remembered by all childhood goldfish enthusiasts, including yours truly, of netting your quarry for temporary residence in a drinking glass while you change the water in his more capacious bowl—a process that can keep Grumpy the Goldfish going for a while, but surely cannot sustain a complex community of aquatic organisms). The concept of an aquarium, on the other hand, rests upon the principle of sustained balance among chemical and ecological components—with plants supplying oxygen to animals, fish eating the growing plants, and snails (or other detritus feeders) scavenging the wastes and gobbling up any algal film that might grow on the glass walls. Western science did not discover the basic chemistry of oxygen, respiration, carbon dioxide, and photosynthesis before the late eighteenth century, so the defining concepts scarcely existed in a usable way before then. The aquarium represents but one of many practical results for this great advance in human knowledge. To quote Shirley Hibberd again: "The Aquarium exemplifies, in an instructive manner, the great balance of compensation which, in nature, preserves the balance of equilibrium in animal and vegetable life."

A few naturalists, before the invention of the aquarium, had managed to keep marine organisms alive for considerable periods in indoor containers—but only with sustained and substantial effort (entrusted to domestic servants, and therefore reflecting another social reality of the times). For example, Sir John Graham Dalyell, a Scottish gentleman with the euphonious title of Sixth Baronet of Binns (and a day job as a barrister to enhance the alliteration), maintained marine animals in cylindrical glass vessels during the early nineteenth century. But he kept only one animal in each jar and had to change the water every day, a job allocated to his porter, who also lugged several gallons of sea water from nearby ocean to baronial home at least three times a week. Sir John did enjoy substantial success. His hardiest specimen, a sea anemone named "Granny," moved into her jar in 1828 and survived until 1887, long outliving the good baronet and several heirs who received this lowly but hardy coelenterate as a legacy that may not have been entirely welcome.

(The history of aquariums has spawned a small but thorough literature. I read this story of Sir John in an excellent article by Philip F. Rehbock, cited in the bibliography to this book. I also benefited from Lynn Barber's general book, published in 1980, *The Heyday of Natural History, 1820–1870.* But I have relied mostly on two primary sources from my personal library: Shirley Hibberd's *Rustic Adornments,* second edition, 1858; and the classic work by one of the greatest Victorian naturalists, *The Aquarium,* by Philip Henry Gosse, second edition, 1856.)

In a similar story, recounted by all major sources on the origin of aquariums, Mrs. Thynne, a lady of means, brought some corals from Torquay to London in 1846 "for the purpose of study and the entertainment of friends" (again quoting Shirley Hibberd). "A stone jar was filled with sea-water; the madrepores [corals] were fixed on a large sponge by means of a needle and thread. They arrived in London safely, and were placed in two glass bowls, and the water changed every other day. But the six gallons of water brought by Mrs. Thynne was now exhausted and must be used again. She here devised means to freshen it for second use." We now switch to Mrs. Thynne's own account, and to another statement about the source of actual work in homes of leisure:

> I thought of having it aerated by pouring it backwards and forwards before an open window, for half or three-quarters of an hour between each time of using it. This was doubtless a fatiguing operation; but I had a little housemaid, who, besides being rather anxious to oblige me, thought it rather an amusement.

In later experiments, Mrs. Thynne did add plants to approximate a natural and sustaining balance, but she never abandoned her practice (or her housemaid's effort) of aeration by hand, and thus never built a truly self-sustaining aquarium: "I regularly placed seaweed in my glass bowls; but as I was afraid that I might not keep the exact balance required, I still had the water refreshed by aeration. I do not know from which, or whether it was from both causes, that my little flock continued to thrive so much, but I seldom had a death."

Interestingly, the key discovery that led to the aquarium of the 1850s did not arise directly from experiments with marine organisms, but by creative transfer from another technology for rustic adornment that had spawned an even more intense craze during the 1840s—the Wardian case for growing and sustaining plants in small, "closely glazed cases." Nathaniel Bagshaw Ward, a London surgeon by profession, began his experiments in the late 1820s. By enclosing plants in an almost airtight glass container—a "closely glazed case" in his terminology—Ward learned how to encourage growth and avoid either desiccation or fouling of the air, all without human input or interference. The moisture transpired by plants during daylight hours would condense on the glass and drip back down to the soil at night. So long as the case remained sufficiently sealed to prevent escape of moisture, but not tight enough to preclude all movement of gases in and out (so that oxygen could be replenished and carbon dioxide siphoned off), the Wardian case could sustain itself for long periods of time.

Dr. Ward's invention provided far more than a pleasant bauble for moral enlightenment in Hibberd's settings of domestic bliss, for the closely glazed case played a key role in Victorian commerce and imperial expansion. Plants in Wardian cases could survive for months at sea, and distant transport became practical for the first time (for species not easily cultivated from seed). In her 1980 book, *The Heyday of Natural History,* Lynn Barber writes:

> The directors of Kew Gardens began to plan even more large-scale movements of plants . . . Literally millions of plants were ferried to and fro in Wardian cases, [and] they eventually succeeded in establishing tea as a cash-crop in India (from China) and rubber in Malaya (from South America), thus adding two valuable new commodities to the British Empire's resources. Kew's Wardian cases were probably one of the best investments the British Government

has ever made, and in fact they were only very recently superseded by the use of polythene bags.

On a humbler, yet massive, scale, Wardian cases also became a fixture in almost every British home of approved taste. Although many kinds of plants could be grown in such cases, a passion for ferns—so spectacular as a social fad that the epidemic even received a latinate description as Pteridomania, or the fern craze—swept Britain in the 1840s. When this mania inevitably subsided, the technology of Wardian cases remained, ready to be utilized for the next enthusiastic bout of rustic adornment—the aquarium craze of the 1850s.

All fads, however brightly they may burn for the moment, seem to run their appointed course in relatively short order. The aquarium craze dominated amateur interest in natural history during the 1850s, but quickly subsided during the next decade. By 1868, another popular naturalist, the Reverend J. G. Wood, could write:

> Some years ago, a complete aquarium mania ran through the country. Every one must needs have an aquarium, either of sea or fresh water, the former being preferred . . . The fashionable lady had magnificent plate-glass aquaria in her drawing room, and the schoolboy managed to keep an aquarium of lesser pretensions in his study . . . The feeling, however, was like a hothouse plant, very luxuriant under artificial conditions, but failing when deprived of external assistance . . . In due course of time, nine out of every ten aquaria were abandoned . . . To all appearance the aquarium fever had run its course, never again to appear, like hundreds of similar epidemics.

Even the most ephemeral episode of public fascination teaches us many lessons about the social and ideological context of all scientific movements. We have already seen how the aquarium craze relied upon chemical discoveries, a philosophical notion about natural balances, a social system that supported a substantial class of domestic servants in wealthy homes, and the development of a technology first exploited in a previous craze for ferns.

Further reading reveals other important ties to political and technological history, most notably the necessary repeal, in 1845, of the heavy tax that had been levied upon glass. Gosse's "how to" book of 1856, *The Aquarium: An Unveiling of the Wonders of the Deep Sea,* exposes the social or technological solution to a number of practical problems that would probably not occur to a casual reader today. How, for example, could an urban enthusiast get sea water for his home aquarium? Gosse advises:

> In London, sea-water may be easily obtained by giving a trifling fee to the master or steward of any of the steamers that ply beyond the mouth of the Thames, charging him to dip it in the clear open sea, beyond the reach of rivers. I have been in the habit of having a twenty gallon cask filled for me, for which I give a couple of shillings.

And how can specimens be safely transported to town with adequate speed? By fast train, of course. Gosse writes:

> The more brief the period during which the specimens are *in tran-situ* the better. Hence they should be always forwarded per mail train, and either be received at the terminus by the owner, or else be directed "To be forwarded immediately by special messenger." The additional expense of this precaution is very small, and it may preserve half the collection from death through long confinement.

Any social movement must illuminate its own time, so we should scarcely be surprised by such enlightenment from the aquarium craze of the 1850s. But what can we say about the even more interesting (and practical) matter of definite and permanent influences extending forward to our own day? Can a movement that trod so transiently (however intensely) on the pathway of history—and then was gone like the wind—leave any lasting imprint upon posterity? In one trivial sense, of course, we can only answer this question affirmatively, for aquariums retain strong popularity in all scales of life—from hokey commercial theme parks, to lofty public museums, to research laboratories throughout the world, to home displays (with an interesting tie to social circumstances, at least in the United States, where cultivation of tropical fishes remains as resolutely working-class as bowling, while the skiing and sailing crowd favors bird watching or African safaris for their natural-history fix).

I take a far greater interest in "invisible" matters usually passing beneath overt notice, because solutions seem so obvious that we do not even acknowledge the existence of a question. Some ways of knowing or seeing seem so blessedly evident, so unambiguously ineluctable, that we assume their universal and automatic practice from time immemorial. Og the caveperson, Artie the australopithecine, even Priscilla the Paleocene primate ancestor, must have used the same devices. But when we can show that such a strategy of thought or sight arose from a recent and specific episode in our actual history, then we obtain our best proofs for the important principle that all knowledge must arise within social contexts—even the most "obvious" factual matter based on direct and simple observation (for one must first ask the right question to secure the proper observation, and all questions emerge from contexts).

Little examples of big principles strike me as most intriguing of all—for I declare my allegiance with several common mottoes proclaiming that God, the devil, or any matter of great pith and moment, lies in the details. I believe that we can identify one of these admittedly small but "obviously" permanent and universal modes of seeing as, instead, a direct legacy of the mid-nineteenth-century aquarium craze, and therefore not much more than one hundred years old as a Western way of knowing.

How shall we draw marine organisms and more-general scenes of underwater communities? The answer to such an inquiry seems so evident that we may wonder why anyone would bother to pose the question at all. We always draw such scenes in their "natural" orientation today: in the "eye-to-eye" or edge-on view, where a human observer sees marine life from within—that is, as if he were underwater with the creatures depicted, and therefore watching them at their own level. Isn't this orientation obviously

In J. J. Scheuchzer's Physica sacra, *marine invertebrates are arranged in "preaquarium" perspective.*

best? After all, we wish to show these creatures as they live, pursuing their ordinary behaviors and interactions. How else could we possibly draw them except from within their own marine environment?

Such a preference may seem both natural and unassailable—and therefore constant and permanent in human practice—but the history of illustration reveals a different and much more interesting story. Until the mid-nineteenth-century, marine organisms were almost always drawn either on top of the waters (for swimming forms, mostly fishes) or thrown up on shore and desiccating on land (for bottom dwellers, mostly invertebrates). These views from above, and from a terrestrial vantage point, had become conventional in the history of art. For example, to invoke the "gold standard" of pre-nineteenth-century illustration for the history of life, consider the engravings for the origin of fishes and marine mollusks in the *Physica sacra* of the Swiss savant J. J. Scheuchzer, published in the 1730s.

This amazing work—the equivalent, for its time, of an elaborate television series with the usual tie-ins from books to coffee mugs—includes 750 gorgeously elaborate, full-page engravings depicting every biblical scene with any plausible implication for natural history. The creation stories of Genesis 1 and 2 provide obvious fodder for an extensive series of illustrations. All marine organisms appear on top, or out of, the waters—that is, from the perspective of a human observer standing on shore. The figure for the creation of mollusks shows clams and snails draped over a rocky arch, or lying on the beach in the foreground, while no organisms at all appear in the background ocean. The creation of marine vertebrates shows a garland of fishes along top and upper side borders (that is, *above* the ocean), while a few swimming whales and fishes partially protrude above the surface, and flying fishes grace the air spaces above!

I can only imagine one reason for a strong convention of such strikingly suboptimal illustration. Artists must then have avoided—or not even been able to conceptualize—the eye-to-eye, within-their-own-environment viewpoint so "naturally" favored today. Illustrators must have eschewed this edge-on orientation because most people had never seen marine organisms

Scheuchzer's marine fishes (seen on the surface of the ocean and as a framing garland) are portrayed from the top-down vantage point of a shore-bound observer.

in such a perspective before the invention of the aquarium, and the craze for maintaining such a display as a rustic adornment in the home converted the formerly inconceivable (because unseen) into a commonplace. Water is usually muddy and largely opaque when in motion. No technology of face masks, diving bells, snorkels, or oxygen tanks existed—and humans do have to come up for air after very short periods of potential observation. The vast majority of Western people (including most professional sailors) couldn't swim, and wouldn't think of immersing themselves voluntarily in marine waters. So where, before the invention of aquariums, would most people ever have seen (or even been inclined to imagine) marine organisms in their own environments? The conventional, if uninformative, view from the shore (and down upon the waters) surely represented the "natural" way of human knowing before aquariums opened a new perspective.

Martin Rudwick, an excellent paleontologist in his early career and now the world's most distinguished historian of geology, first made me aware of this interesting change in the history of illustration, and the probable inspiration provided by the invention of aquariums. In his remarkable book on the history of drawings for prehistoric life (*Scenes from Deep Time,* 1992), Rudwick noted that virtually all early illustrations depict marine organisms exclusively as assemblages desiccating on shore—quite a limit for learning about past communities and environments, especially when you realize that most of life's history featured marine organisms only! Rudwick writes:

> Most scenes from deep time . . . portrayed ordinary marine organisms as having been washed up on a shore, in the foreground of a landscape seen unproblematically from a human viewpoint. In this respect, they simply continued the established pictorial convention . . . In effect, the aquatic world from which most fossils were derived was depicted only from the outside, from the subaerial-world to which a time-travelling human observer could more plausibly have had access . . . This suggests how difficult it may have been for the public . . . and perhaps for most of the geologists too, to

imagine a viewpoint that was not only prehuman but also subaqueous—at least until mid-century, when the famous aquarium craze made the underwater world generally accessible for the first time.

I don't mean to exaggerate the exclusivity of this theme. The eye-to-eye view is not *that* hard to imagine, even if one has never seen marine life in this orientation—and fishbowls did provide some simplified hints. Thus, one occasionally encounters the "modern" view in old illustrations. (The earliest I have seen comes from a sixteenth-century German book on military tactics, and shows a soldier—or should I say a marine—stealthily walking along a lake bottom to access an enemy ship and drill some holes for a sinking. The figure shows a few fishes swimming in the water, but in a very stiff and clearly subsidiary role.)

But Rudwick is surely correct in noting the rarity of such drawings— and he also points out that occasional exceptions usually involve irregular or humorous purposes, while the same artists then used the conventional on-shore view in textbooks and other standard sources. For example, in 1830 and long before the aquarium craze, Henry de la Beche, the first director of the British Geological Survey and a skilled illustrator as well, made a famous drawing of Mesozoic marine life in Dorset—from the "modern" eye-to-eye perspective. He printed this figure as his contribution to a campaign designed to raise money for Mary Anning, the celebrated fossil collector who had become impoverished. But when de la Beche, only two years later, published figures of the same ichthyosaurs and plesiosaurs in a popular textbook, he drew these animals either on shore or on top of the waters.

I have informally monitored this theme in my historical readings during the past five years—and I can affirm Rudwick's claim that the "natural" edge-on view did not become at all "obvious" until the aquarium provided a venue for ordinary human observation. Moreover, since all inventions experience some "lag time" before general acceptance, I have also noted that eye-to-eye marine views do not predominate during the aquarium craze of the 1850s, but only achieve preferred status during the next two decades. To

In the first edition of Louis Figuier's The Earth Before the Flood *(1863), a lithograph depicts Devonian sea creatures cast up on the beach.*

cite two examples of reluctance to abandon old conventions, Shirley Hibberd (in 1858) does show several figures of aquariums from the side. But nearly all Hibberd's drawings, while presenting a side view through glass, take the perspective of an observer looking down upon an aquarium from above, not directly from the side (and level with his fishes). Moreover, Hibberd's decorative drawings for the first page of each chapter continue to promote the desiccating shore-bound view, as illustrated by the grotto of invertebrates gracing chapter one on the "marine aquarium."

In a striking example (cited by Rudwick as well), the immensely popular French naturalist Louis Figuier—the Carl Sagan or David Attenborough of his day—published the first major book of chronologically ordered scenes for each period of life's history (*La terre avant le déluge,* or *The Earth Before the Flood*). His lithographer, Edouard Riou, also worked for Jules Verne (among others) and was the most celebrated illustrator of popular science in his time. In the first edition of 1863, Riou drew all marine creatures in positions of death and desiccation on shore. He retained these

Illustrating marine animals of the Carboniferous period, a postaquarium perspective is evident in Louis Figuier's fourth edition (1865).

figures in later printings, but added, in the fourth edition of 1865, a much more informative drawing of Carboniferous fishes and marine invertebrates in the newly familiar edge-on aquarium view.

Very little comes easily to our poor, benighted species (the first creature, after all, to experiment with the novel evolutionary inventions of self-conscious philosophy and art). Even the most "obvious," "accurate," and "natural" style of thinking or drawing must be regulated by history and won by struggle. Solutions must therefore arise within a social context and record the complex interactions of mind and environment that define the possibility of human improvement. To end with a parody on a familiar text, we only learned the "natural" way to see marine life when the invention of aquariums permitted us to see through glass clearly, and to examine a brave old world face to face.

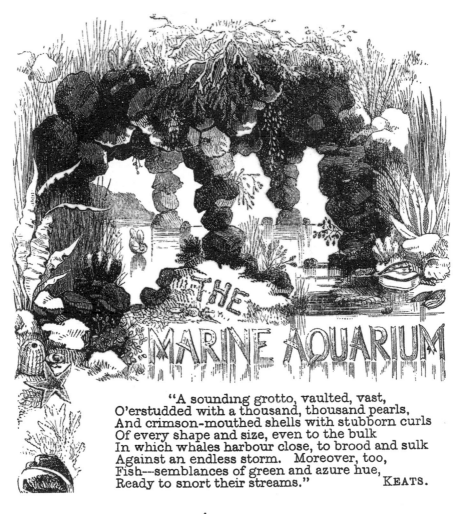

THE MARINE AQUARIUM

"A sounding grotto, vaulted, vast,
O'erstudded with a thousand, thousand pearls,
And crimson-mouthed shells with stubborn curls
Of every shape and size, even to the bulk
In which whales harbour close, to brood and sulk
Against an endless storm.   Moreover, too,
Fish—semblances of green and azure hue,
Ready to snort their streams."   KEATS.

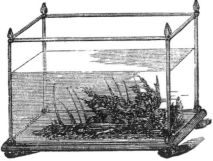

A chapter heading in Rustic Adornments *reflects the old style of marine illustration, as does the top-down view of the aquarium itself.*

II
BIOGRAPHIES IN
EVOLUTION

4

THE CLAM STRIPPED BARE
BY HER NATURALISTS,
EVEN

In Benjamin Britten's operatic setting of Henry James's *Turn of the Screw,* the boy Miles sings a little ditty to his governess during their Latin lesson:

> *Malo: I would rather be*
> *Malo: in an apple tree*
> *Malo: than a naughty boy*
> *Malo: in adversity*

Britten embodies all the fear and mystery of James's eerie story in setting this doggerel as a searing and plaintive lament that then cycles throughout the opera, emerging at the very end, but this time intoned by the Governess as Miles lies dead on the stage. Britten's device works well because Miles's text is so insipid (yet at the same time so expressive of his fears about personal evil). The English doggerel scans properly, rhymes, and makes sense, but the pedagogic joke lies in the fact that each of the four English lines (made of several words) can be fully translated by the single Latin word *malo* (the first person singular of the verb *malle,* to prefer; the ablative of the noun *malus,* an apple tree; etc.).

Miles's poem, in fact, belongs to a venerable genre of crutches devised to make children love their Latin—obviously an ancient problem for teachers to overcome. Latin versions of various children's classics—*Winnie Ille Pu* most prominent among them—represent our most conspicuous modern effort toward the same end.

But children of the generation just before mine often encountered a much more pungent spur to their diligent study—namely, sex. Several men of my father's generation have told me that they applied themselves earnestly to the ancient tongue because some neighborhood kid always had access to his parents' copy of Krafft-Ebing's *Psychopathia Sexualis*—that great late-nineteenth-century compendium of case studies in every conceivable kind of sexual peculiarity (so graphically expressed that even Mr. Justice Stewart would recognize the genre). The main text had long before been translated into English—but, following Krafft-Ebing's own expressed wishes, all the juicy case studies remained only in Latin!

I missed all the fun. I never learned the kiddie mnemonics because I only studied Latin in graduate school. And I never relished the sexual prod because Krafft-Ebing's case studies had made their way into English before my prurient years. Thus I greatly enjoyed a little belated amusement last week when I finally got some ribald pleasure out of all that graduate-school effort. But I wasn't reading Krafft-Ebing. I was studying the 1771 treatise on mollusks, *Fundamenta testaceologiae* (never translated from its Latin original), by none other than Carolus Linnaeus.

Yes, we are discussing clams—though Linnaeus seems to be talking about the sexual anatomy of women. Linnaeus's treatise begins in the stolidly conventional mode of most taxonomies. He states that he will classify mollusks by their shells (so often prized by naturalists), rather than the animals within (biologically better, but rarely collected). He then makes a primary division into *Cochleae* (basically snails, with a few other single-shelled creatures thrown in, including scaphopods and even an errant worm tube or two), and *Conchae* (basically clams, or bivalves, but also housing multivalved mollusks like chitons and a few creatures that don't belong at all by genealogy, including brachiopods and barnacles).

He then, still in conventional fashion, provides a list of technical terms for the parts of shells—and he begins his compendium for clams with one of the most remarkable paragraphs in the history of systematics. He regards the hinge between the two valves *(cardo)* as a defining character, and he then writes: *Protuberantiae insigniores extra cardinem vocantur Nates*—or "the notable protuberances above the hinge are called buttocks." He then names all the adjacent parts for every prominent feature of sexual anatomy in human females—*ut metaphora continuetur* ("so that the metaphor may be continued"). Clams have a *hymen* (the flexible ligament connecting the two valves at top), *vulva, labia,* and *pubes* culminating in a *mons veneris* (various features at the top of the shell behind the umbos—our modern term for Linnaeus's buttocks); and, in front of the umbos, an *anus.*

Linnaeus's forced rationale for these terms can best be grasped from the picture he presents in illustration—for a species that he named *Venus dione,* no doubt as a fitting illustration for his terminology. The picture, with my

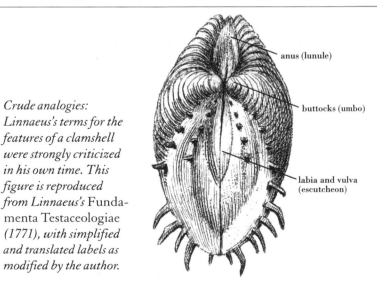

anus (lunule)

buttocks (umbo)

labia and vulva
(escutcheon)

Crude analogies: Linnaeus's terms for the features of a clamshell were strongly criticized in his own time. This figure is reproduced from Linnaeus's Fundamenta Testaceologiae *(1771), with simplified and translated labels as modified by the author.*

added labels, shows the full crudity of Linnaeus's supposed joke—for his terms record a complex analogy (not overly far-fetched, one must admit, in purely visual appearance) between a clamshell viewed from the top and the standard sleazy pose of pornography: a woman with legs widely spread and sexual parts viewed straight on, with buttocks surrounding two areas of external genitalia and anus.

Linnaeus was socially conservative and rather prudish. He did not, for example, allow his four daughters to study French, for fear that they would then learn the liberal values of that enlightened land. But his taxonomic systems and his writings reveal the sexual focus that so often accompanies personalities of such overwhelming vigor and force. Linnaeus based his most celebrated work, his new classification of plants, on what he called the "sexual system" (see the final section of essays in my previous book, *Dinosaur in a Haystack*). This scheme tends to be dry and functional, and not at all salacious—for the sexual system defines most orders by the numbers and sizes of stamens and pistils, the male and female organs of flowers. Basically, you just have to count—and Linnaeus's system became all the rage for ease of application, not for titillation. But Linnaeus did follow out the metaphorical implications of his definitions. He referred to fertilization as an act of marriage, and he designated stamens and pistils as husbands and wives. Flower petals turn into bridal beds, and infertile stamens become eunuchs, guarding the wife (pistil) for other fertile stamens. Linnaeus wrote, in an essay of 1729:

> The flowers' leaves . . . serve as bridal beds which the creator has so gloriously arranged, adorned with such noble bed curtains and perfumed with so many soft scents that the bridegroom with his bride might there celebrate their nuptials with so much the greater solemnity.

But these botanical images rank as sweet Arcadian romances compared with the overt salaciousness of his terminology for the parts of clams.

Consequently, Linnaeus took a great deal of contemporary flak about his names for the top side of clamshells.

In 1776, a good year for reform, an obscure English naturalist, who lived a shadowy and troubled life (as we shall see), lit into the great master for his licentious malfeasance. In the preface to his *Elements of Conchology: or, An Introduction to the Knowledge of Shells,* the author wrote:

> One subject, however I shall insist upon; that is to explode the Linnaean obscenity in his characters of the Bivalves; . . . Science should be chaste and delicate. Ribaldry at times has been passed for wit; but Linnaeus alone passes it for terms of science. His merit in this part of natural history is, in my opinion, much debased thereby.

Late in the book, as the author reaches his chapter on clams, his fury has not abated. This time he advances the explicit argument that Linnaeus's terminology makes natural history appear hostile to females, thus discouraging intellectual women from pursuing one of the few areas then relatively open to study by all people:

> I am the more desirous of fixing technical names, as the unjustifiable and very indecent terms used by Linnaeus in his Bivalves may meet their deserved fate, by being exploded with indignation; for
>
> *Immodest words admit of no defense,*
> *And want of decency is want of sense.*
>
> These my terms being adopted, will render descriptions proper, intelligible, and decent; by which the science may become useful, easy, and adapted to all capacities, and to both sexes.

(I originally assumed that the author's heroic couplet must represent a quotation from Alexander Pope, but my trusty *Bartlett's* tells me that the lines belong to an obscure character named Wentworth Dillon, Earl of

Roscommon [1633–1685]—a solid name indeed for poetic utterances of such unexampled propriety.)

The author of this book, Emmanuel Mendes da Costa, was a Sephardic Jew of Portuguese origin. He was born in London in 1717 and died there, in his lodgings on the Strand, in 1791. Although Mendes da Costa became one of England's most respected naturalists on the undefinable borderline between amateur and professional status; although he maintained voluminous correspondence (much apparently preserved in the British Museum) with many of Europe's greatest naturalists and with most major players in the widespread network of British amateurs, his name has almost entirely disappeared from the historical record—except for two lovely books that frequently appear on the antiquarian market: the 1776 treatise on conchology, and his 1757 work, titled *The Natural History of Fossils* (I base this essay largely on these two works). I may well have missed some secondary sources,[1] but I was unable to find anything about Mendes da Costa's life and

1. Indeed, I did miss the only modern article published about Mendes da Costa, an excellent and thorough account by P. J. P. Whitehead in 1977 (see bibliography for details). I regret this oversight, both for my failure to honor Whitehead's extensive work, and for the considerable extra time and labor I therefore spent in locating (from scratch and for a second time) the rare and sketchy sources that we both used as a basis for our articles. (Scholarship is truly a collective and accumulative exercise in this crucial sense, as I failed both Whitehead and myself. To be sure, Whitehead published his account in a small and highly technical journal not included in most standard bibliographies. But, no excuses. My sources on Mendes da Costa were far more obscure and far more difficult to locate!)

Whitehead's article treats a quite different subject from my focus (the true authorship of Mendes da Costa's book on conchology, which Whitehead determines to be, indeed, da Costa's own work, and not that of other contenders). But he also includes an interesting account of Mendes da Costa's life. We reach very much the same conclusions, but Whitehead's work does resolve one issue in the original version of my essay (here corrected). I had been unable to find enough evidence to convince me of Mendes da Costa's guilt in charges laid against him by the Royal Society in 1767, and leading to his several years in prison. But Whitehead proves that Mendes da Costa was indeed guilty as charged, though his deep regret and subsequent rehabilitation also seem well documented.

In one crucial theme of my essay, Whitehead ably affirms my central claim that Mendes da Costa has—curiously, given his unique and fascinating status—elicited remarkably little attention from later historians and scientists. Whitehead writes: "Da Costa has so far received only

works beyond a column entry in the British *Dictionary of National Biography,* a few bits and pieces in early-nineteenth-century volumes of *The Gentleman's Magazine,* and, fortunately, about fifty pages of his fascinating letters reproduced in volume four of an 1822 series by John Nichols titled *Illustrations of the Literary History of the Eighteenth Century Consisting of Authentic Memoirs and Original Letters of Eminent Persons.*

I regard this erasure of Mendes da Costa as most unfortunate for at least two reasons: because he must have led a fascinating life, and because his history illustrates several social and scientific issues of general importance, including the role of amateurs in natural history, and the status of Jews in eighteenth-century England. I shall focus this essay on another theme in the category of general messages well displayed: Mendes da Costa's role as a leading collector at the crux of a defining transition in natural history. For he practiced at both ends of the passage from primary concern for weird specimens and star items (the biggest, the most colorful)—the *summum bonum* of the seventeenth-century baroque age, as embodied in the tradition of constructing natural-history collections as *Wunderkammern,* or chambers of curiosities—to the eighteenth-century passion for order in the classical world of the Enlightenment. Linnaeus's new system acted as a prerequisite for Darwin's revised explanation of causes. But the older love of oddity continued to fan public enthusiasm (and still does so, quite appropriately, today).

Mendes da Costa was an ordinary man in the midst of this great transition. And ordinary people often record patterns of history with maximal fidelity and interest—for Mendes da Costa made no attempt to innovate on a grand scale, and he therefore becomes a standard for his age. In Linnaeus we grasp the thrust of change. By studying Mendes da Costa, we can best understand the fixed beliefs, the impact of novelty introduced by innovators, and, particularly, the intellectual impediments that his age posed to better

---

brief biographical treatment although, like many of his colleagues, he was an avid letter-writer and his carefully preserved correspondence (over two thousand letters) still survives." May I hope that some aspiring Ph.D. candidate in the history of science reads this footnote and finds a superb thesis topic thereby? Then Whitehead's article and this essay will truly not have been in vain.

comprehension of the natural world. We must learn to view these impediments with proper sympathy—not in the old style of condescension for an intellectual childhood to compare with our stunning maturity, but as a set of consistent and powerful beliefs, well suited to the culture of another time, and held by reasonable people with raw intellects at least as good as ours. If we can achieve such fairness and equipoise, the history of science will become the greatest of all scholarly adventures—and also the most utilitarian, for the foibles of the past can only help us to grasp our own equally constraining present prejudices.

So ordinary, and yet so different! Partly for reasons of self-definition, but mostly from the contumely of others, Jews have lived apart within most Western nations, often with cruel restrictions attached (see chapter 13). Emmanuel Mendes da Costa grew up in Britain at an interesting time, probably more favorable than most, for his people.

Jews had dwelled in England from the Norman Conquest until their expulsion under Edward I in 1290. After banishment from Spain in 1492, and from Portugal in 1497, Sephardic Jews (named from the Hebrew word for Spain) dispersed widely, but still could not settle in England. Some small communities of *conversos* or *Marranos* (officially converted Jews, but many still practicing their old religion secretly) lived in England from time to time, but when Shakespeare wrote the *The Merchant of Venice,* and created the anti-Semitic character of Shylock, no openly practicing Jews inhabited England. A new group of *Marranos* began to enter Britain from Rouen in the 1630s. This community, hoping for more toleration from Oliver Cromwell's Protectorate than from the previous monarchy, petitioned for the right to practice their religion openly—and their plea received favorable action in 1656. The restored monarchy of 1660 did not rescind the permission, and a few Jews therefore continued their tenuous tenancy. They could not, for example, engage in retail trade in London until 1822, and could not sit in Parliament until 1858 (Disraeli was a Christian convert).

As part of this history, very few Jews inhabited England in Mendes da Costa's day—about two thousand Sephardim by the end of the eighteenth-century, and perhaps somewhat more Ashkanazim, or Jews of German and eastern European origin. In a potentially prejudiced society, just a few folks from an alien culture may appear exotic and fascinating, rather than threatening and despised—and the rarity of Jews seemed to work in Mendes da Costa's favor, as he often encountered philo-Semitism among his noble and gentleman correspondents.

Emmanuel Mendes da Costa was trained in law, but chose to devote himself to natural history. He built a fine collection and published several articles, leading to his election as a fellow of the Royal Society (England's premier association of scientists) in 1747, and to the Society of Antiquities in 1751. But his troubled and shadowy side also surfaced amid his successes. The *Dictionary of National Biography* remarks: "Although he early obtained the reputation of being one of the best fossilologists at his time . . . his life appears to have been a continual struggle with adversity." He was imprisoned for debt in 1754. After his release the next year, he began to prepare, and finally published in 1757, his major treatise *The Natural History of Fossils*.

Mendes da Costa received his biggest opportunity in 1763 when he became clerk of the Royal Society, in charge of their collections and library, then in a state of neglect and disrepair. He wrote to a friend in September 1763:

> I immediately proceeded to work, but such was the state of the said libraries and museum, that I am inclined to think the Augean stable was but a type of them [a reference to Hercules' most unpleasant labor, far more taxing than killing the Lernaean Hydra, of clearing thirty years of manure from the stable of Augeas, King of Elis] . . . After many weeks' work, amidst the repeated curses of myriads of spiders and other vermin, who had held peaceable possession for a long series of years, I accomplished, so that, thank God, now both libraries and museum are accessible, and in a state fit to be consulted by the curious.

Nonetheless, Mendes da Costa took great joy in the good fortune of his new job. He wrote to another friend: "Whenever you come to town, pray let me see you. Our Museum here, I assure you, has many fine things, and our library is very numerous and scientific. I am very happy in my places, and henceforward my whole life will be devoted to study." But four years later, in December 1767, he was dismissed for "various acts of dishonesty," arrested at the suit of the Society, and committed to the King's Bench prison, where he remained until 1772. His library and collections were also seized and sold at auction.

Mendes da Costa continued his work under confinement, aided by the support and patronage of several well-placed friends. On January 3, 1770, he wrote to a Dr. Francis Nicholls:

> I received your much esteemed letter, which honors me with an invitation to your house at Epsom, to review some fine minerals you have lately collected in Cornwall . . . But I am so unfortunate at present as not to be able to embrace the much desired and respected offer you make me; as I am under confinement in this King's Bench . . . However, the Almighty who had afflicted me with the confinement, has through His mercies granted me the call of my reason, and I apply myself as much as ever, and assiduously to my studies.

Four years later, Nicholls still remembered, and wrote:

> It is with pleasure I hear you are restored to liberty and philosophy; and that you should like to see my collection of Cornish fossils . . . My son will come down next Sunday morning; so, if you will be at his house in Lincoln's Inn-fields by nine, he will bring you down, and render your journey less tedious.

Mendes da Costa soldiered on, writing increasingly more obsequious letters in hopes of selling specimens or delivering lectures for a fee. His worst debacle and embarrassment occurred in 1774, when his petition to deliver a series of lectures at Oxford was not only summarily rejected, but scorned with the overt politeness often used by powerful patricians before they squish a plebeian favor-seeker like an insect. Apparently, Mendes da Costa made the mistake of submitting a formal proposal, when he needed to work through channels and secure the verbal permission of the Vice-Chancellor (the boss of the university). (I doubt, in any case, whether a Jewish jailbird would have received such sanction under any circumstances at the time.) Mendes da Costa did finally prevail upon a professor to visit the Vice-Chancellor, who promptly spurned the idea: "The course of lectures proposed to be read by Mr. Da Costa could not be read here with propriety. I hope the disappointment will sit easy upon Mr. Da Costa." In fact, the rejection weighed most heavily, as Mendes da Costa wrote to the professor:

> I am very certain my attempt has not succeeded by means of some unfriendly and sinister misrepresentations, as well as through mismanagement on my side, for want of proper advice how to proceed. I unluckily had not a friend who chose by a single line to set me right, or inform me what to do . . . Thus left forlorn, absent from the scene of action, and ignorant how to proceed, I became shipwrecked, and my hopes were blasted.

But Mendes da Costa never gave up. He published his conchology book in 1776 to good notices, rebuilt his collections, kept up his correspondences, and died in reasonable honor.

Throughout this various life, one theme keeps circulating in constancy: Mendes da Costa's Judaism, and the fascination thus inspired among his philo-Semitic Anglican friends. Mendes da Costa must have become a semi-official source on Jewish matters for the British intelligentsia, at a time when very few English Jews could have traveled in these circles, neatly balancing enough assimilation to find acceptance with sufficient practice of Judaism to be regarded as authentically exotic. In 1751, a physician inquired of him "whether there is extant any where a print or drawing, or any account of the dress and arms of a Jewish soldier, or whether the Jewish Soldiers did not wear the same dress as the Roman Soldiers." Mendes da Costa replied that he did not know, since Jewish sources do not permit representation of human images:

> In regard to any drawing, *etc.,* we never permitted any in our books, apparel, *etc.,* it not being agreeable to the religion . . . yet I do not find that drawings were at all used in books, *etc.,* even by the Greeks and Romans.

In 1747, Mendes da Costa had to forgo a ducal invitation in order to celebrate the High Holidays at home. But His (philo-Semitic) Grace understood very well and hastened to reassure poor Mendes da Costa, who greatly feared that he had offended a high potential patron. The Duke's secretary wrote:

> His Grace is very sorry the duties of your religion, which every good man is well attached to, prevent your coming hither just at this time . . . The Duke being the most humane and the best man living, you need be in no difficulty about your eating, here being all sorts of fish, and every day the greatest variety of what you may feed on without breach of the Law of Moses, unless the lobsters of Chichester should be a temptation by which a weaker man might be seduced.

In 1766, Mendes da Costa hears of some Hebrew inscriptions at Canterbury, and he writes to an acquaintance there:

> In a MS of Dr. Plot's dated June 10, 1674, I find this notice: 'Antient inscriptions on ruinous buildings—such as the Hebrew exquisitely written on the old walls of the Castle of Canterbury.' Is there such a Hebrew inscription now extant? If there is, can a copy be procured? or can I have permission to employ some Jew (of Canterbury) to copy it, and decypher it.

His friend passed the request to an Anglican scholar who knew Hebrew, for a Jew would not be able to gain access. This scholar wrote directly to Mendes da Costa:

> The Hebrew inscription you inquire after was written on the walls of one of the stone stair-cases in the old castle at Canterbury, in the 13th century, by the captive Jews, during their imprisonment there, and contained some few versicles of the Psalms . . . It is, I do suppose, no very difficult task to get admittance to this inscription, by any gentleman of the County, or one supported by proper recommendations; but I think they would make great objections to admit a stranger and a Jew to search for it.

Amid these signs of both philo- and anti-Semitism, we may also complete the gamut with purely benevolent ignorance. A correspondent writes to Mendes da Costa in 1755, offering payment in goods for services rendered in identifying specimens of natural history: "It is said by most people that Yorkshire hams are very much admired, and if you should think so, will send you some up." The editor appends a telling footnote at this point: "Mr. Knowlton seems not to have recollected that he was writing to a Jew."

Most telling for the history of science, we learn from the correspondence how Mendes da Costa stood on the cusp of a transition between two great sequential worlds of natural history—from the baroque passion for gathering oddities, to the classical urge to order and classify in a single comprehensive system. The quest for oddities certainly emerges in this offer from a correspondent on December 9, 1749:

> I have some natural curiosities to present you with . . . I have the tooth, or tusk, of the sea-lion, . . . part of a young elephant's tooth, in the section of which is an iron bullet, which had been shot into it when younger, and the ivory grown over the bullet; a hair-ball, found in the stomach of a calf; and a fossil or two; which shall all find their way to your Cabinet if you think them worthy a place in it.

But Mendes da Costa's own requests mostly record his concern for completion and order. He asks a Jewish friend in Bath to collect as may kinds of fossils as possible, and to send them to his local coffeehouse—a striking example of different services rendered by public places before the days of home mail delivery:

> In regard to fossils, see if you could get me any *ammonitae,* or snake-stones, as they are vulgarly called, as also impressions of plants on a kind of coal slate, which abound in the collieries. At Lincomb and Walcot are stone quarries which afford very fine petrifactions of shells, *etc.* Could you procure any of these things, and send me a box full directed to the Bank Coffee-house, I shall cheerfully repay all charges whatsoever.

Mr. Schomberg, presumably a German Jew by name, knew what he wanted in return: "Send me a small pot (of about three or four pounds) of sour-crout, . . . and take care it is well secured, so as not to be broke."

Over and over again, Mendes da Costa begs his correspondents to pack carefully and label properly:

> Of whatever is collected, let each specimen be carefully wrapped up and numbered, and a catalogue made with answerable numbers to each specimen, wherein specify what it is, what is vulgarly called, where found, whether in plenty or rare, at what depths, among what other fossil bodies, and all the other curious particulars you can be informed of to elucidate the natural history of them. I beg pardon for troubling you thus, but I am greatly obliged to you for this great piece of friendship.

Mendes da Costa's most assiduous correspondence dates from the 1740s and 1750s as he collects, and beseeches friends to procure, as many "fossils" as possible for his forthcoming comprehensive treatise *The Natural History of Fossils.* Following contemporary custom, Mendes da Costa secured subscriptions to this book before publication (a favored fund-raising device for expensive works). His substantial list, published after the preface to his work, includes six bishops and five lords, further testifying to his acceptability among the Anglican upper classes. (Mr. Joseph Harris, "assay master of the Tower," also signed on for a copy.)

Mendes da Costa published a first substantial volume of his treatise in 1757, but never proceeded any further. This work is, nonetheless, his masterpiece—and a superb example of the passion for fully comprehensive order that so motivated eighteenth-century natural history. In Mendes da Costa's time, "fossil" (from the past tense of the Latin *fodere*, to dig up) referred to any natural object taken from the earth, not only to the remains of organisms. In fact, rocks and minerals were the quintessential "fossils" of eighteenth-century terminology, since they belong to the mineral kingdom—as natural products of the earth—whereas remains of ancient plants and animals must be introduced into rocks from the animal and vegetable kingdoms. Thus, the bones, shells, and leaves, now exclusively honored with the term *fossil*, were called "extraneous fossils" in the eighteenth century—while rocks and minerals represented essential fossils. Mendes da Costa therefore intended to cover all products of geology in his treatise—rocks, minerals, and remains of ancient organisms. But his first and only volume did not proceed beyond rocks and earths. (If he had completed the full design, and had followed the usual classifications of his time, he would probably have written a second volume, on minerals and crystals, and a third, on remains of organisms.)

Mendes da Costa may have been disgusted at Linnaeus's sexual terminology for clams, but the great Swedish naturalist still reigned as the prince of order. We know that Linnaeus's binomial system persists in largely unaltered form as the basis for giving scientific names to organisms today. This system has served us so well (despite recognized limitations) that we forget the fallacies of its original overextended application—an amnesia also abetted by our tendency to view the history of science as a list of growing successes, with errors buried into a conventional metaphor: "the ash-heap of history."

But Linnaeus's original application of binomial nomenclature suffered under the common eighteenth-century fallacy of grandiosity in attempting to encompass all possible kinds of natural diversity in one system of classification. For Linnaeus didn't apply his binomial system only to plants and animals (where the procedure has always worked well, for reasons discussed below), but also, in essentially unaltered form, to minerals, and even, in his *Genera morborum* of 1763, to diseases, which he grouped by their symptoms into classes, orders, and genera.

The Linnaean system implies a definite geometry for the ordering of objects, and therefore only works when the causes of order produce results conforming with this geometry. Consider two essential properties: First, the Linnaean system is hierarchical. The basic units (species) are grouped into genera, genera into families, families into orders, and so on. This scheme implies a treelike topology with the largest unit (say, the kingdom Animalia) as a single trunk; middle units as large boughs attached to the trunk (phyla like Arthropoda and Chordata); smaller units as branches emerging from the boughs (classes like Mammalia and Aves joined to the chordate bough); and, finally, basic units as twigs growing from the branches (species like *Homo sapiens* and *Gorilla gorilla* attached to the primate branch of the mammalian bough of the animal trunk). This topology correctly represents a system of objects produced by branching evolution, with continuous divergence and no suturing together of separately formed branches. Since the history of life operates by this geometry, the Linnaean system works splendidly for classifying organisms.

Second, basic units must be discrete and definable, not smoothly intergrading and constantly joining. Since organic species are independent and stable units (after the brief geological moment of their branching origin), the Linnaean system also functions ideally for complex, sexually reproducing organisms.

But the same reasons that allow the Linnaean system to classify fossil organisms so well also guarantee an inapplicability in principle to the two categories of the mineral kingdom that Mendes da Costa's age also called "fossils"—minerals and rocks. Minerals and their crystals have definite chemical formulae, and aggregate by simple physical rules. Their relative similarities are not genealogical, and their order cannot therefore be expressed by a treelike geometry. Moreover, mineral "species" are not discrete entities composed of genealogically related individuals in historical continuity. Cambrian quartz, at half a billion years of age, does not differ from Pleistocene quartz separately made in a geological yesterday.

Rocks and soils, composed of mineral grains and the products of their erosion, fail the Linnaean requirements for an even more fundamental reason. Rocks and soils form a broad continuum of fully intergrading compositions. We cannot identify discrete species of granites, marbles, or chalks. Granites, for example, are composed of quartz, two kinds of feldspar, and a dark mineral like biotite or hornblende—and all these constituents can be mixed and matched like house paint into any desired composition.

Nonetheless, Mendes da Costa, as a devotee of the classical passion for order, struggled to do as the master Linnaeus had enjoined—to make every object of nature fit the binomial system, thereby bringing all phenomena into one grand style of order. Thus, in *The Natural History of Fossils,* Mendes da Costa presents a Linnaean classification of earths and stones into species, genera, and other categories now used only for organisms. His great treatise has a wonderfully archaic ring today because he seems to treat objects of the mineral kingdom as a collection of organisms, and sets, as his highest goal for this branch of geology, a grouping of rocks to match a listing of beetles. I am particularly fond of his *Natural History of Fossils* because this treatise, more than any other work written in English, records a short episode expressing one of the grand false starts in the history of natural science—and nothing can be quite so informative and instructive as a truly juicy mistake.

Consider Mendes da Costa's classification of earths and stones. He does not use Linnaeus's own names for categories, but he follows the same basic procedure. Linnaeus's hierarchy included four levels (we have since added several more): class, order, genus, species. Mendes da Costa uses six: series, chapter, genus, section, member, and species. At the highest level, he divides his geological objects into two series: *earths* and *stones.* Following the

Linnaean principle (and a long history of Western thought traceable to Aristotle), he provides a definition of categories based on a *fundamentum divisionis,* or basic criterion of difference. Earths are "fossils not inflammable, but divisible and diffusible, tho' not soluble in water"—whereas stones have the same properties, but are not divisible and diffusible.

He separates the first series, *earths,* into seven genera within three chapters. Chapter 1, defined as "naturally moist, of a firm texture and which have a smoothness like that of unctuous bodies," includes three genera—boles *(Bolus),* clays *(Argilla),* and marles *(Marga).* Chapter 2 ("naturally dry or harsh, rough to the touch and of a looser texture") encompasses two genera—chalks *(Creta)* and ochres *(Ochra).* Finally, Chapter 3 ("naturally and essentially compound, and never found in the state of pure earth") also includes two genera—loams *(Terra miscella)* and molds *(Humus).*

The second series, *stones,* includes nine genera in four chapters, based on interesting criteria that we would now regard as partly superficial and partly on the mark for wrong reasons. The four chapters comprise (1) stratified rocks made of grit (basically sandstones, divided by Mendes da Costa into two genera for finely stratified with many horizonal bedding planes versus massive and blocky); (2) stratified rocks without grit and homogeneous (divided into two genera—basically limestones and slates in modern parlance—by the same criterion of massive versus thinly bedded); (3) marbles (separated as much for their importance to human arts as for any other reason); and (4) crystalline rocks, divided largely by the size of mineral grains into basalts and other finely crystalline rocks, granites, and porphyries.

Mendes da Costa presents an interesting rationale—though ultimately flawed—for why a system that has worked so well for organisms should render equal service to inorganic geology. "It has been by pursuing such natural and simple methods as these, that botany has so eminently raised her head above her sister sciences," he writes, acknowledging Linnaeus's greatest success.

Mendes da Costa recognizes the differences in formation between organic and inorganic objects, but he follows a common scientific conceit in thinking that a uniform system of classification will nonetheless suffice for all: he will engage only in humble and accurate description, not in fanciful theorizing. Differences in causality cease to matter when we cite only the pristine factuality of objective appearances: "I have been very cautious not to indulge a speculative fancy in forming hypotheses or systems, the bodies being simply described, according to the appearances which they exhibit to the senses."

Mendes da Costa then declares success because he has managed to balance, in a single system of compromises, all the competing schemes of his contemporaries. Such a "golden mean" must yield optimality. Mendes da Costa argues that he has achieved two great balances in his system—first, by finding the "right" number of basic species as a compromise between "splitters" who love to make fine distinctions, and "lumpers" who search for essences and tend to unite objects in their quest for fundamental properties. (The terminology of splitting and lumping belongs to the twentieth century, and the implied dichotomy cannot express all subtleties of this contrast, but the struggle between joiners and dividers has pervaded the history of taxonomy.) Mendes da Costa writes: "I have endeavored to reduce this study, hitherto deficient in respect of method, to a regular science, and in the attempt have been careful neither to multiply the species, nor lessen their number, unnecessarily."

In a second balance, Mendes da Costa tries to unify the two disparate criteria then used to form systems for rocks and minerals—the efforts of his British compatriot John Woodward to base distinctions on observation of overt properties, both exterior and interior ("a method of arrangement founded on the growth, structure, and texture of fossils"); and continental systems based on "essential" properties discovered by chemical experiment (for example, a threefold division according to various changes produced by fire into *calcarii,* for rocks calcined, or turned to lime [limestones and marbles, for example], *apyri* for those unaffected [asbestos and others], and *vitrificentes* for those vitrified to glass [quartz and other silicates]). Mendes da Costa tries to bring all systems together by making primary divisions with observable properties (Woodward's system), and then applying experimental and chemical results for refinements:

> I have attentively examined the Woodwardian and Wallerian [continental] systems, and, finding them defective, have presumed to form a new one from the principles of both. I have endeavored to arrange fossils, not only according to their growth, texture, and structure, but also their principles and qualities, as discovered by fire, and acid menstrua, *etc.* And in this way, I am confident that all the known fossils may be accurately distinguished; whereas, to attempt it by any one system hitherto followed, must occasion a strange confusion.

But Mendes da Costa's efforts had to fail in principle because the causes and properties of rocks violate the requirements of Linnaean geometry. Following the two central fallacies discussed earlier in this essay, Mendes da Costa could neither identify clear species, nor form distinct categories in a world of complete intergradation. Biological species are natural populations, distinct by historical continuity and current interaction, and unable to interbreed with others. Rock "species" are nondiscrete and intergrading. Ultimately, Mendes da Costa just joined specimens that looked "enough alike"—a sure formula for endless bickering among specialists, for no two will ever agree. For example, he castigates his two great masters, Linnaeus and Wallerius, for designating too few species in the genus *Marmor* (marbles):

> Wallerius, in his *Mineralogy,* and Linnaeus, in his *Systema Naturae,* are extremely confused in regard to this genus of fossils; the former had divided all the marbles into only three species; *viz.* of uniform, variegated, and what he calls figured marbles . . . the latter has even made them all only varieties of one species; on which I cannot but make this observation, that it is a pity the learned should apply their studies rather to perplex science, than to elucidate it, and instruct mankind.

The Clam Stripped Bare by Her Naturalists, Even

(I am no Freudian, but one certainly glimpses Oepidal feelings in Mendes da Costa's complex attitude toward Linnaeus—basing his life's work on the taxonomic system of his intellectual father, but then losing no opportunity to razz the master for moral turpitude on a range of issues from the terminology of clams to the number of species of marbles.)

But Mendes da Costa cannot claim final certainty for his own divisions of *Marmor.* He designates eighty-one species, far more than for any other genus of rocks, and clearly to recognize the human utility of different colors and patterns, not because nature has fabricated discrete and discoverable "basic kinds."

Equal difficulty and frustration dogged Mendes da Costa's effort to establish the higher divisions of rocks and earths—for he encountered complete intergradation between his genera as often as tolerable separation. For example, he frankly states his difficulty in dividing boles from clays, finally admitting that only convention dictates the separation:

> Several authors do not make a distinct genus of the boles, but rank them among the clays; indeed very essential characteristics are wanting to make them different genera, for only the extreme fineness of the particles of the boles is the cause of their being not so ductile or viscid as the clays, insomuch that speaking with propriety, they are only to be accounted very fine clays; I have, however, made them separate genera, as custom hath authorized it.

The human mind, with arrogance and fragility intermixed, loves to construct grand and overarching theories—a fault perhaps encountered more often in our theological than in our scientific endeavors. But solutions often require the humbler, superficially less noble, and effectively opposite task of making proper divisions into different categories of meaning and causation. For only then can we build toward generality on a firmer substrate, and without feet sculpted of the genus *Argilla.* Mendes da Costa, following Linnaeus, tried to bring all nature into one grand system of classification, but principles appropriate for the branching of organisms do not suit the continuities of rocks and earths. How ironic, ultimately, that a system doomed by too much togetherness should embody the best work of Emmanuel Mendes da Costa, the only Jewish naturalist of note in eighteenth-century Britain—a man from a culture then defined by separation from a majority committed to the parochial notion that Anglican theology represented the one true and only way.

To heap irony upon irony, and to end with a return to the beginning, Mendes da Costa did understand the general principle that now leads us, with such confidence and justice, to classify rocks differently from organisms. I omitted one line in his critique of Linnaeus's sexual terminology for bivalves (as indicated by the ellipses on page 81), for these words cite a technical objection, rather than the moral argument then under discussion. In this line, now restored, Mendes da Costa rejects Linnaeus's bivalve names "not only for their licentiousness, but also that they are in no ways the parts expressed." How simple, and how correct! The top of a clam is not the bottom of a person—and supposed visual similarities can only be misleading. Different terms should be used, lest people be lulled into false suppositions about meaningful or causal likeness. Similarly, rocks and organisms require different systems of classification to acknowledge their disparate modes of production—by timeless chemistry versus singular genealogy, by laws of nature versus contingencies of history.

But all human beings belong to a single fragile species, a biological unity too much divided by the worst emotional traits of our common nature. Separate the stones from the snakes, but let Emmanuel Mendes da Costa, a Jewish stranger in a strange land, shake hands with his Mad King George—for then, perhaps, "they shall not hurt nor destroy in all my holy mountain," made, no doubt, of the genus *Granita,* from the Italian for grain, to signify all the bits and pieces of diverse minerals that come together to form this hard rock of unity.

5

DARWIN'S AMERICAN
SOULMATE:
A BIRD'S-EYE VIEW

I HAVE LONG CONSIDERED ABRAHAM LINCOLN AS CHARLES DARWIN'S American soulmate—for they were born on the same day, February 12, 1809. But perhaps the accidents of joint beginnings should not define a concept of such intimacy. If soulmates must be linked more tightly by their active choices, then Darwin's American alter ego can only be his fellow scientist James Dwight Dana (1813–1895)—geologist, biologist, longtime professor at Yale, and surely America's preeminent indigenous natural historian of the nineteenth century. (Louis Agassiz, the other obvious contender, was born in Switzerland and did his important scientific work in Europe before coming to Harvard University in the late 1840s.)

Dana and Darwin never met personally—though they both expressed a warm desire to do so in their numerous letters. But their careers and interests ran in intricate, almost eerily parallel courses. Both men had their scientific baptism in a long sea voyage around the world—Darwin on the

Beagle from 1831 to 1836, and Dana on the Wilkes expedition of 1838–1842, young America's greatest international scientific journey, dispatched primarily to assess whaling prospects in the southern oceans. Both men, similarly inspired by their travels, then built their scientific careers on the same two subjects.

Darwin's first scientific book, published in 1842, presented a correct theory on the origin of coral atolls by subsidence of a central island, with continued upgrowth of living coral at the edges. Dana also became fascinated with corals when he visited Pacific reefs. In 1839, while ashore in Sydney, Dana chanced upon a description of Darwin's ideas in a local newspaper. With this inspiration, Dana produced the other great nineteenth-century work on coral reefs—*Corals and Coral Islands*—substantially supporting Darwin's "subsidence theory," but based on far more extensive observation than Darwin had been able to make. In the preface to the second edition of his book on coral reefs, Darwin wrote:

> The first edition of this book appeared in 1842, and since then only one important work on the same subject has appeared, namely . . . by Professor Dana . . . It has also afforded me the highest satisfaction to find that he accepts the fundamental proposition that lagoon-islands or atolls, and barrier-reefs, have been formed during periods of subsidence.

In a second parallel, both Darwin and Dana devoted their major technical work in zoology to the taxonomy of the same group of organisms—the crustacean arthropods. Darwin published four volumes between 1851 and 1854 on the oddest of crustacea, the barnacles. Dana based fourteen years of research on specimens collected by the Wilkes expedition, and published his finest work in 1852—two volumes on the taxonomy of crustaceans.

In fact, Darwin's barnacles inspired the first personal contact between the two men in 1849, when Darwin wrote to ask if he might borrow specimens collected by Dana's expedition. (Thus, the striking similarity of their careers had developed without any direct mutual encouragement.) Darwin wrote, rather formally:

> I hope that you will forgive the liberty I take in addressing you . . . in order to beg, if it lies in your power, assistance . . . It is my earnest wish to make my monograph as perfect as I can. Can you lend me any species collected during your great expedition?

Dana replied, warmly but sadly, that he would be personally delighted to do so, but had neither possession nor authority over the specimens. Darwin understood, and wrote Dana a long letter of praise for his work, noting: "You cannot imagine how much gratified I have been that you have to a certain extent agreed with my coral island notions."

A warm epistolary friendship ensued. Darwin wrote three years later, in 1852:

> You ask whether I shall ever come to the United States. I can assure you that no tour whatever could be half so interesting to me, but with my large family I do not suppose that I shall ever leave home. It would be a real pleasure for me to make your personal acquaintance.

(Darwin knew himself well. After sailing around the world and returning to England, he never again left his island home; he never even crossed the English Channel!)

The next year, Darwin enthused over Dana's volumes on crustacea, just published:

> If you had done nothing else whatever, it would have been a *magnum opus* for life. Forgive my presuming to estimate your labors, but when I think that this work has followed your *Corals* and your *Geology,* I am really lost in astonishment at what you have done in mental labor. And then, besides the labor, so much originality in all your works!

Despite this effusion of warmth and mutual support, Dana and Darwin inevitably parted company on the great issue that would define their time (and ours)—evolution itself. As I shall document later, Dana did eventually succumb to the inevitable in the mid-1870s, but his late support for evolution always remained strictly limited, clearly grudging, and only pursued as a necessary compromise to save as much as possible of his unaltered world-view. But Dana remained a staunch, if always cordial, opponent of evolution throughout the great debates of the 1860s, the defining decade after Darwin published the *Origin of Species* in 1859.

Darwin sent Dana a copy of the first edition, but Dana's health had broken, and he did not read the book until 1863. Nonetheless, Dana could not avoid the issue in publishing the first edition of his most famous work in 1862—the *Manual of Geology*. While stating his opposition in the book, Dana also thought that he owed his epistolary friend a personal explanation. So he wrote to Darwin on February 5, 1863: "I hope that ere this you have the copy of the *Geology* (and without any charge of expense, as was my intention). I have still to report your book unread; for my head has all it can now do in my college duties."

Dana then spelled out his major objections, all paleontological, in a series of three points. Dana's arguments show that he based his opposition upon his personal definition of evolution as a necessarily progressive and gradual process. If evolution were valid, Dana claimed, then the history of life would have to proceed by slow and steady transformation from simple to complex forms in each lineage. Dana then listed his objections:

1. The absence, in the great majority of cases, of those transitions by small differences required by such a theory . . .
2. The fact of the commencement of types in some cases by their higher groups of species instead of the lower . . .
3. The fact that with the transitions in the strata and formations, the exterminations of species often cut the threads of genera, families, and tribes . . . and yet the threads have been started again in new species.

Dana had cited a set of highly traditional objections—lack of transitional forms, first geological appearance of advanced rather than primitive members of a lineage, and mass extinctions—and Darwin rebutted them all (not entirely successfully, as later history would show) by arguing that a woefully imperfect fossil record had rendered a truly gradual and progressive history of life in this deceptive manner. Darwin did feel that he owed Dana a personal reply, and he wrote back with his usual rationale on February 20, 1863—just two weeks after Dana had dated his letter, and not bad for sea transport of mail during the Civil War:

> With respect to the change of species, I fully admit your objections are perfectly valid. I have noticed them . . . I admit the same if the geological record is not excessively imperfect.

Then, in the only hint of rancor that I have ever detected in their correspondence, Darwin upbraids Dana for stating these objections without reading his book—though he quickly falls back into his usual geniality, and also assures his friend that he only felt aggrieved because Dana's opinion carried so much weight:

> As my book has been lately somewhat attended to [lovely British understatement], perhaps it would have been better if, when you condemned all such views, you had stated that you had not been able yet to read it. But pray do not suppose that I think for one instant that, with your strong and slowly acquired convictions and immense knowledge, you could have been converted. The utmost that I could have hoped would have been that you might possibly have been here or there staggered.

The personal and intellectual drama of Darwin and Dana provides the main subject for this essay, but I also write to illustrate a broader theme in the lives of scholars and the nature of science: the integrative power of worldviews (the positive side), and their hold as conceptual locks upon major innovation (the negative side). I will argue that Dana was not benighted, stupid, or particularly stubborn. Rather, he maintained a consistent, well-articulated, and clearly coherent theory of God and life—a worldview that just didn't contain any logical space for a Darwinian concept of naturalistic evolution. One does not (and probably should not) surrender the system of a lifetime for one apparently errant bit of information. If we can do so at all, we will surrender such sources of succor and consistency with the same feelings that we experience when we leave a natal home or a first great love—slowly, sadly, prayerfully, and, above all, with deep affection and respect.

The issue of when to hold firm as suggestions of factual collapse accumulate, and when to plunge with abandon into the breach, defines the most interesting and important dilemma of intellectual autobiography—for this decision defines the borderline between competence and genius, or between sensibility and crankiness. In some crucial sense, the geniuses of history are people who know when to plunge and how to create the instruments of successful assault and replacement. But we must also remember that probably 99 percent of personal plunges fling potential heroes into whirlpools of error and erasure from history. Still, the message of these failures should not inspire calls for sticking with the tried and true at all costs—or else the earth would still occupy the center of a small universe, and people would still be the manufactured incarnations of a divinely ordained perfection. Most people, including the most polished intellectuals of every generation, never dare to take the plunge. This phenomenon of psychology and society produced the old cliché, usually attributed to the nineteenth-century German physicist Ernst Mach, that new theories only triumph fully when the old guard dies off.

Dana's conservative, and ultimately superseded, worldview rested upon two central convictions that made Darwinian evolution impossible (not so much factually wrong as literally inconceivable within such a system). First, Dana was as pure a Platonist as nineteenth-century biology ever produced. He based his zoological ideas firmly on the old concept of "type"—an idealized form for each group of animals, with variation among individuals of a species as accidental departures from ideal propriety, and variation among species as organized sequences obeying "laws of form" that expressed the thoughts and plans of divinity. (Agassiz, Dana's equally Platonic colleague, declared taxonomy to be the highest science, for each species represents a divine idea made flesh, and the arrangement of species therefore expresses the structure of divine thought. By understanding the system of order among species, we therefore achieve our closest insight into the character of God's mind.)

Second, Dana viewed the entire geological history of the earth and life as one long, coherent, and heroic story with a moral—a tale of inexorable progress, expressed in both physical and biological history, and leading, inevitably and purposefully, to God's final goal of a species imbued with sufficient consciousness to glorify His name and works. The physical earth, according to Dana, developed through time with the same gradualistic progress that defined life's history. By following three major trends—the emergence of more and more land from the sea, the purification of the atmosphere, and progressive global cooling with consequent increase in climatic diversity by formation of zones from poles to tropics—the earth became more and more suited to the successively higher forms of life that God created in each new episode of progress. For inhabitants of the land must be ranked "higher" than denizens of the sea; pure air inspires healthy complexity (slithering reptiles in murky swamps versus sinewy mammals on the bright plains); and cooler climates require such advances as warm-blooded metabolism. Dana wrote in his *Manual of Geology:*

> Thus the prevalence of waters involved inferiority of species. The increase of land, the gradual purification of the atmosphere, and the cooling of the globe, prepared the way for the higher species.

Lest anyone be tempted to read the sequence of successive creations, each with increasing excellence, in an evolutionary manner à la Darwin, Dana always took pains to state that such a history could only record the direct actions of a loving God with a goal in mind. Dana wrote in 1856:

The whole plan of creation had evident reference to Man as the end and crown of the animal kingdom, and to the present cool condi-tion of the globe as, therefore, its most exalted state. It is hence obvious, that progression in the earth from a warmer to a cooler condition, necessarily involved progression from the lower to the higher races . . . The earlier races were of lower types, not because the Creative Hand was weak, but for the reason that the times, that is the temperature and condition of the globe, were just fitted, in each case, for the races produced, and the progress of the plan of creation, correspondingly, required it . . . The development of the plan of creation . . . was in accordance with the law of . . . progress from the simple to the complex, from comprehensive unity to multiplicity through successive individualizations.

Dana buttressed each of his two controlling ideas with a definite biological theory of his own construction. In attempting to explain why and how the history of life should feature a vector of progress as the fundamental thrust of created change through time, Dana invented an influential notion that he named "cephalization," or increasing domination of the head. Crustaceans are segmented animals, and their classification largely depends upon the form and number of appendages on the head segments. (In ancestral arthropods, each body segment presumably bore a pair of legs. In later evolution, these appendages often become concentrated and specialized. Many paired organs of crustaceans and other modern arthropods—antennae, mouthparts, swimming paddles, claspers used as external genitalia—evolved from modified legs. Insect and crustacean mouthparts, especially when greatly magnified in TV nature specials, look particularly weird and, to use the modern vernacular, yucky, because we correctly perceive them as a bunch of little legs waving about—and legs just don't protrude from mouths on a "proper" vertebrate model!) Dana, under the controlling influence of his progressivist worldview, therefore chose to arrange the diversity of modern crustaceans in sequences of primitive to advanced, as defined by complexity and domination of the head and its appendages. He then extended this principle of "cephalization" to all of animal life.

Dana first published his theory of cephalization in the mid-1850s, before learning about Darwin's ideas. But he wrote his four major articles on the subject between 1863 and 1866 (all published in the *American Journal of Science,* a periodical that he edited).

I find the theory of cephalization wonderfully fascinating and a bit mad. If I could write a volume, rather than an essay, on this subject, I would gladly discuss each of Dana's sixteen criteria (many with several subdivisions) for the supposedly objective measure of precise degrees in organic "highness" and "lowness" based on the character of the head and its relative domination of the body. I would also demonstrate the self-serving (for his worldview, not his persona) and "fluid" nature of these criteria—scarcely objective in any reasonable sense, but obviously constructed in order to validate Dana's a priori desires, and almost wantonly fudged or changed whenever an apparent exception arose.

For example, one criterion states that progress can be measured by position of the head and brain along the body axis—the farther forward the better. Thus, Dana regards whales as "low" mammals because they have so much mouth and snout in front of the brain. But as soon as he gets stuck because a group he wants to regard as primitive happens to place a brain right up front, he just shifts his criteria. For example, he wants to place millipedes and centipedes below insects, but these many-legged arthropods situate their head and brain right at the front end. So he pronounces this particular head as weak, although located in the "right" place—as measured by minimal domination over the rest of the body because "lowly" legs proliferate behind. Dana writes: "The head is here strictly at the anterior extremity; but the cephalic force has so feeble control, that the joints multiply behind." Dana even admits, in several passages, that his criteria may be inconsistent, but he then argues that such complexity requires ever more subtle interpretations, and ever more experienced interpreters, to make the system work!

Each of Dana's papers attempts a summary definition of cephalization as the dominant motor of life's progressive history. His "last fling" on the subject, a final article in 1876 after a ten-year hiatus, offers this account:

> In the low, there is, usually, large size and strength behind, an elongation of the whole structure, and a low degree of compactness in the parts before and behind; in the high, there is a relatively shorter and more compacted structure, a more forward distribution of the muscular forces or arrangements, and a better head; and the progress in grade . . . is progress along lines from the former condition toward the latter, that is, progress in the strength, perfection and dominance of the anterior or cephalic extremity; in a word, it is progress in cephalization.

Although Dana may have developed this concept of cephalization from his work on crustacea, and although he claimed an objective status for his theory, any thorough reading will reveal that he cobbled his theory together in order to proclaim the message that controlled his worldview: the centrality and domination of human life in the cosmos. Why choose cephalization as the criterion of progress, except as a way of exalting our own big head, located right at the top, with anterior limbs freed from the lowly task of locomotion and dedicated instead to the head's service (as I type this essay with my fingers, but only curl my toes uselessly under my chair)? Dana wrote:

> While all other Mammals have both the anterior and posterior limbs as organs of locomotion, in Man the anterior are transferred from the locomotive to the cephalic series. They serve the purposes of the head, and are not for locomotion. The cephalization of the body—that is, the subordination of its members and structure to head uses—so variously exemplified in the animal kingdom, here reaches its extreme limit. Man, in this, stands alone among Mammals.

To buttress his second dominating idea of created type, Dana developed an idiosyncratic taxonomy within a popular pre-Darwinian genre that evolution would soon render incoherent—a numerological system, with a fixed number of subgroups in each larger group. Dana favored either two or four subgroups as the key to correct classification.

Historians of taxonomy have often argued—quite incorrectly—that the development of evolution inspired little change in the structure of classification, for the order that had once been attributed to God could easily be shifted to evolution without any alteration of content. This claim supposedly carries the message that theories should be viewed as mental constructions sitting lightly upon nature—or, to put the argument another way, that nature's evident factuality must be rendered in the same manner by any chosen mode of explanation. In either formulation, the role of theory, or worldview, gets demoted and the proper balance between inside theory and outside nature becomes distorted by deemphasizing the internal aspect of a truly intricate and equal pairing.

But this common claim must be rejected. Evolution made a world of difference in classification. The major groups may have retained their definitions (arthropods are arthropods, and vertebrates vertebrates, whether created by God or developed by evolution), but a hundred other significant details had to change because the geometry of evolution differs fundamentally from the structure of created systems. Numerological schemes like Dana's disappeared forever as soon as Darwinism triumphed—and most modern taxonomists don't even know that such systems ever existed (and therefore fail to appreciate the power of evolution to alter the practice of their profession), because numerology suffered such a complete and sudden death. If God made all species, and their order reflects the nature of his thought, then why not search for an arcane numerological system that might embody divine wisdom? But if organisms are tied together by genealogy on an evolutionary tree of life, then success or failure becomes a question of contingent history, and no rationale for fixed numbers of subgroups within groups can possibly be devised.

When Dana worked by twos, he divided each group into a "typical" class defining the essence, and a "hemitypical" class specifying a departure.

Thus, for example, he regards terrestriality as typical for vertebrates (don't ask me why, for fishes came first—but I do suspect an a priori desire to define the group containing humans as typical). The twofold division of vertebrates therefore contrasts Tetrapoda (all terrestrial forms) as typical, with Pisces (fishes) as hemitypical.

When Dana worked by four, he specified three degrees of typicality in descending order—alphatypic, betatypic, and gammatypic—with a fourth group as a true departure named "degenerative." In this version, mammals are alphatypic (for standing tall), birds betatypic (lovely creatures, but not on defining terra firma), reptiles gammatypic (as slitherers on the ground), and fishes degenerative (as living in the "wrong" place for vertebrates).

In a revealing article, titled "Thoughts on Species" (including mineralogical as well as biological entities), and published in 1857 as Darwin composed his magnum opus, Dana defended his numerological system as embodying an unchangeable, Platonic, and universal truth:

> Fixed numbers, definite in value and defiant of all destroying powers, are well known to characterize nature from its basement to its top-stone . . . The universe is not only based on mathematics, but on finite determinate numbers in the very natures of all its elemental forces.

He even argued that the human soul requires fixed numbers, both to avoid despondency by perceiving order, and to adore God even more. (This passage also expresses Dana's hostility to any notion of graded evolutionary transition between groups.)

> Were these units capable of blending with one another indefinitely, they would no longer be units, and species could not be recognized. The system of life would be a maze of complexities; and whatever its grandeur to a being that could comprehend the infinite, it would be unintelligible chaos to man. The very beauties that might charm the soul would tend to engender hopeless despair in the thoughtful

mind, instead of supplying his aspirations with eternal and ever-expanding truth. It would be to man the temple of nature fused over its whole surface and through its structure, without a line the mind could measure or comprehend.

Darwin's candid reaction to Dana's two theories of cephalization and numerological taxonomy provides a striking illustration of the unbridgeable discordance between their worldviews. On February 17, 1863, in the interval between Dana's key letter on evolution and Darwin's emotional reply, Darwin also wrote to his guru and confidant Charles Lyell about his unhappiness with Dana's paper on classification of mammals by principles of cephalization and numerical order. Darwin, with his usual insight, nails Dana for constructing a system whose absurdity would be apparent to anyone not firmly committed to ranking humans as the crown of creation. Dana's whole scheme, Darwin correctly notes, becomes one long, forced rationale for human centrality:

> The same post that brought the enclosed brought Dana's pamphlet on the same subject. The whole seems to me utterly wild. If there had not been the foregone wish to separate man, I can never believe that Dana or any one would have relied on so small a distinction as grown man not using fore-limbs for locomotion, seeing that monkeys use their limbs in all other respects for the same purpose as man. To carry on analogous principles . . . from crustacea to the classification of mammals seems to me madness. Who would dream of making a fundamental distinction in birds, from fore-limbs not being used at all in some birds, or used as fins in the penguin, and for flight in other birds?

(Darwin's valid complaint provides another example of how theory can control the arrangement of nature. Dana used the functional status of the fore-limb as a key character for making major taxonomic divisions among birds because, under his theory of cephalization, the head controls use of the fore-limbs, and their status therefore defines the degree of domination by the head. Darwin, in the last sentence of his letter to Lyell, regards such a scheme as absurd because different uses of forelimbs, in his evolutionary version of life, must be interpreted as immediate adaptations to varying modes of life, not as fundamental divisions on the genealogical tree of birds.)

As a chief ingredient in the mythology of science, the accumulation of objective facts supposedly controls the history of conceptual change—as logical and self-effacing scientists bow before the dictates of nature and willingly change their views to accommodate the growth of empirical knowledge. The paradigm for such an idealistic notion remains Huxley's famous remark about "a beautiful theory, killed by a nasty, ugly little fact." But single facts almost never slay worldviews, at least not right away (and properly so, for the majority of deeply anomalous observations turn out to be wrong; every fact about the revolution of the earth can be paired with a hundred claimed observations for cold fusion, perpetual motion, or the transmutation of gold).

Rather, at least for a first approach, anomalous facts get incorporated into existing theories, often with a bit of forced stretching to be sure, but usually with decent fit because most worldviews contain considerable flexibility. (How else could they last so long, or be so recalcitrant to overthrow?) The best test for the power of a worldview to order and interpret facts—and, therefore, an excellent illustration of the fascinating and intricate interaction between theory and data in science—arises when someone discovers an absolutely pristine and unanticipated bit of novel information. Fortunately, I practice a profession—paleontology—particularly well supplied with superb test cases, for nothing can be quite so out-of-the-blue as a newly discovered fossil. Therefore, if we consider a discovery that, in modern hindsight, points unambiguously toward the validity of a new worldview, we achieve our ideal test case: if everyone bows to the fact immediately, and accepts the implied reconstruction of nature, then Huxley's dictum triumphs. But if most members of the old guard manage to embrace the new fact comfortably enough within their conventional worldview, then major changes of theory in science require a more complex push involving social context as well as factual impetus.

In the early 1860s, as Darwin and Dana debated evolution in their letters, the best possible example of an unanticipated fact burst upon the scene—the discovery of *Archaeopteryx,* not only the oldest bird but also, apparently, beautifully intermediate between reptiles and birds in its retention of teeth, reduced coating of feathers, and basically reptilian anatomy. Score one knockout blow for evolution.

Darwin, of course, read the discovery in exactly this sensible light. He wrote to Dana on January 7, 1863:

> The fossil bird with the long tail and fingers to its wings . . . is by far the greatest prodigy of recent times. This is a great case for me, as no group was so isolated as birds; and it shows how little we knew what lived during former times.

But wait. The old fighter rises at the count of nine. He circles back; he feints, he bides his time; the bell rings. He rests and recoups; and he comes out fighting for the next round. In November 1863, Dana published his reply to *Archaeopteryx* in an article titled "On Parallel Relations of the Classes of Vertebrates, and on Some Characteristics of the Reptilian Birds." *Archaeopteryx,* he proclaimed, provides no evidence for evolution, but becomes instead the best possible discovery for validating his own creationist numerology of classification based on cephalization!

Since we are primates, and primates are visual animals, we often epitomize our worldviews in iconographic form. And nothing can be quite so powerful as a picture for summarizing and solidifying a view of life. In his article, Dana presents the classification of vertebrates as a picture, and thereby upholds a crucial role for *Archaeopteryx* in completing the geometry of divine numerology. As his picture shows (see the following figure), Dana wishes to classify each of the three terrestrial classes—mammals, birds, and reptiles—into his customary twofold division of typical and hemitypical. In each case, the hemitypical group should point toward the inferior class

I.

A. Typical Mammals.

B. Hemitypic Mammals,

or OÖTICOIDS.

II.

A. Typical Birds

B. Hemitypic Birds,

or ERPETOIDS.

III.

A. Typical or True Reptiles.

B. Hemitypic Reptiles,

or AMPHIBIANS.

IV.

A. Hemitypic Fishes,

or SELACHIANS.

B. Hemitypic Fishes,

or GANOIDS.

C. Typical Fishes,

or TELEOSTS.

This 1863 diagram by James Dwight Dana epitomizes his "creationist system of numerology."

---

below. Mammals, of course, stand on top. Ordinary placentals form their typical group, while marsupials and the egg-laying monotremes (duck-billed platypus and echidna) build the hemitypical group below. He calls these hemitypical mammals "ooticoids"—and they clearly point to birds and reptiles in their laying of eggs.

Reptiles also come in twos—the typical snakes, turtles, lizards, and others of that ilk; and the hemitypical frogs and salamanders among the amphibians. (We now classify amphibians as a separate class, but taxonomists of Dana's day often put all terrestrial coldblooded creatures together into an expanded class Reptilia.) As hemitypic egg-laying mammals pointed to the reptilian class below, so also do hemitypic amphibians point to the fishes below in their initial aquatic phase of tadpole life.

But what about birds? Now Dana encounters a problem, and a threat to the numerical beauty of his system. Flying birds are clearly typical, but what birds can be called hemitypical? Hemitypes must point to the class below—in this case, again to fishes. One might label flightless ostriches and emus as hemitypical, but in what possible way can these creatures be pointing to fishes? On the contrary, their terrestrial life seems to point upward to mammals (or at least sideways to reptiles), and therefore threatens the entire system. Imagine, then, Dana's delight in the discovery of *Archaeopteryx*—for he could argue that this fossil's retention of teeth leads downward to the hemitypic shark, which, as Mack the Knife so pointedly noted, has "pretty teeth, dear, and he shows them pearly white." Almost gleefully, therefore, Dana portrayed *Archaeopteryx,* the supposedly final messenger of evolutionary truth, as the salvation of his own numerological system of creation! For *Archaeopteryx*— "erpetoids" in Dana's terminology—stands forth as the missing hemitypic bird, and Dana's system becomes healed and whole. He wrote:

> The discovery of the Reptilian Birds has brought the general law to view, that, among the four classes of Vertebrates, ordinarily received, each, excepting the lowest, consists of, first a grand typical division, embracing the majority of its species, and secondly, an inferior or hemitypic division, intermediate between the typical and the class or classes below.

Dana actually drew two arguments against evolution from his new icon of vertebrate classification. First, as noted above, *Archaeopteryx* completed a numerical geometry that could only arise by divine intent and imposition. The organization of fishes provides a second anti-Darwinian argument. Typical fishes are teleosts, the bony fishes that include almost all modern species. But fishes also include two hemitypic divisions—with the crucial difference that fish hemitypes point *upward* toward the higher terrestrial vertebrates rather than downward. The hemitypic sharks (Selachians on Dana's chart) point to the hemitypic *Archaeopteryx* and then up to typical birds; and the hemitypical lungfishes (Ganoids on Dana's chart) point up to the hemitypical amphibians and thence to the typical reptiles.

The whole system is therefore rounded and complete within itself. The upper divisions point down through their hemitypical groups; while the lower divisions point up through their own hemitypes. What else but a static and created order could be so self-contained and self-defining? Dana concludes, with explicit refutation of Darwin:

> It is plain from the preceding that the subkingdom of Vertebrates, instead of tailing off into the Invertebrates, has well-pronounced limits below, and is complete within itself . . . We find in the facts no support for the Darwinian hypothesis with regard to the origin of the system of life.

This scheme may be madness, as Darwin might have said, but it is surely divine madness.

Historians, like most decent people, tend to be patriotic. Dana represented America's best, and who wants to saddle him with a reputation as an old fogey and holdout against the truth of evolution? Several scholarly articles have therefore focused upon Dana's belated, strictly minimal, and reluctant "conversion" to evolution—as first stated in the 1874 edition of his *Manual of Geology* and, two years later, in his last paper on cephalization. Rather defensively, Dana now holds (in the 1876 article on cephalization) that evolution may have become the preferred mode of change, but that progress by cephalization still marks the result. Dana now seems to say, "I was right about what happens, but perhaps not about how it happens. What happens is more important anyway. Pattern reveals divine intent; mechanism is only a means to an end." In his own words:

> Whatever the types of structure in course of development, there was also a general subordination in the changes to the principle of cephalization . . . These views may hold whatever be the true method of evolution. The method by repeated creations through communications of Divine power to nature should be subordinated, as much as any other, to molecular law and all laws of growth; for molecular law is the profoundest expression of the Divine will . . . But the present state of science favors the view of progress through the derivation of species from species, with few occasions for Divine

intervention. If then there has been derivation of species from species, we may believe that all actual struggles and rivalries among animals, leading to a "survival of the fittest," must tend, as in Man, to progress in cephalization.

But we should not cite such grudging passages as Dana's last hurrah and ultimate redemption—though traditional interpretations have followed this route. Dana made his minimal move toward evolution in order to preserve as much as possible of his crumbling system, not as a zealous, born-again crusader who had finally seen the light. By owning evolution as a mechanism, Dana could preserve his deeper convictions about progress by cephalization.

The "heroic" approach does great disservice to Dana's powerful intellect, and perpetuates a silly doctrine of validation by redemption in late conversion to a current truth (almost like an apostate Christian who reconciles with Jesus on his deathbed and expires in grace, despite the ignominy of his former existence). Dana's last hurrah for evolution was a little blip, not the definition of his scientific life. We should honor and respect him for the power of his lifelong view, ably and honorably defended over decades, though now judged wanting. Surely, in science, it is no sin to be wrong for good reasons.

If we dismiss those scientists now judged wrong, only valuing them if they eventually saw the light, we will miss a grand opportunity to address one of the most elusive and portentous questions in scholarly life. What is the nature of genius; why, among brilliant people, do some make revolutions and others die in the dust of concepts whose time had begun to pass in their own day? What is the crucial difference between Darwin's transcendent greatness and Dana's merely ordinary greatness? (Ordinary greatness is not an oxymoronic concept, but a definition of leadership in old guards throughout history.)

I do not know the answer to this question of questions, but we can surely specify a key ingredient. Somehow, for some reason of psyche or quirk of mind, some impetus of social life or some drive of temperament, Darwin was driven to challenge, to be fearless in bringing down an intellectual universe, to be joyful in trying out each thrilling and lovely bit of furniture in a reconstructed world. Dana, for other properties of the same attributes, could not, or dared not, abandon the traditional hope and succor of centuries: Rock of ages, cleft for me; let me hide myself in thee.

Consider, in closing, how the two men treated Plato, the greatest of all intellectual Gods. Dana simply revered his name and his concept of a permanent realm of idealized perfection. Darwin delighted in challenging the master—in showing how simply, and how elegantly, the new evolutionary view could interpret and explain some of the great mysteries and arcana of the ages. Just one comment, privately penned in one of Darwin's youthful notebooks, after he returned to London on the *Beagle,* captures this fundamental difference between Darwin's flexibility and Dana's immobility. With one line, Darwin cuts through two thousand years of traditional interpretation for innate concepts of the human brain. They are not, he nearly shouts for joy, manifestations of Platonic absolutes transmitted from the ideal realm of archetypes, but simple inheritances from our past:

> Plato says in *Phaedo* that our "imaginary ideas" arise from the preexistence of the soul, and are not derivable from experience. Read monkeys for preexistence!

6

A SEAHORSE FOR

ALL RACES

RICHARD OWEN, ENGLAND'S GREATEST ANATOMIST, AWAITED WITH KEEN anticipation the monthly installments of Charles Dickens's latest novel, *Our Mutual Friend*. Owen needed no special reason to join his countrymen in reading the serialized work of England's most beloved writer. But Owen did have a personal stake in the new book, for Dickens had shaped the character of Mrs. Podsnap for his scientific friend: "A fine woman for Professor Owen, quantity of bone, neck and nostrils like a rocking-horse, hard features, majestic head-dress in which Podsnap has hung golden offerings."

Our Mutual Friend appeared in full form in 1865. In the same year, perhaps in specific gratitude, but perhaps only to acknowledge their general friendship, Owen inscribed a copy of his newly published *Memoir on the Gorilla* to "Charles Dickens, Esq. from his friend the Author." I regard my ownership of this copy as a rare and precious privilege. Dickens made no annotations, but a bookplate on the cover, presumably inserted as a come-on for a sale after Dickens's death in 1870, does prove that Owen's friend kept and shelved the book: "From the library of Charles Dickens, Gadshill Place, June 1870." The friendship of Owen and Dickens blossomed within that bastion of Victorian connectivity among males of good breeding or accomplishment (sometimes even both): club life. They met most frequently at the Athenaeum—the major London club for intellectuals, and including both Darwin and Huxley as members. The Athenaeum still exists and still excludes women from several of its spaces. Traditions and memories, both good and bad, die hard. I was once shown the very spot on the main staircase where Dickens and Thackeray almost came to blows.

In our current consciousness, gorillas have become familiar, if continually fascinating. But in Owen's day, mystery and novelty increased the allure of these largest apes. Chimpanzees had been known for more than a century (the London physician Edward Tyson had written a classic monograph on chimp anatomy in 1699), while Dutch ships had brought orangutans back from Indonesian colonies. But the gorilla, though featured in numerous legends, did not prove its existence to scientists until 1846, when Thomas Savage, an American missionary, obtained some skulls in Gabon. Owen, who had published many papers on the anatomy of other apes and monkeys, narrowly lost the race for priority in identifying and naming the gorilla, when the French anatomist Isidore Geoffroy Saint-Hilaire and the American physician Jeffries Wyman barely beat him into print.

But Owen, as chief of natural history at the British Museum, had maximal access to new specimens. In 1851, he received the first complete skeleton to reach England, followed in 1858 by a nearly full-grown male preserved in spirits. In 1861, the Museum purchased a collection of skins for mounting and exhibition, including females, males, and juveniles shot by the explorer Paul B. du Chaillu. Owen, therefore, possessed both the skills and the material to become the first great scientific expert on gorillas—and he accepted the challenge in many publications, culminating in his 1865 monograph.

Owen had the skin, muscles, and bones, but knowledge of behavior and ecology still depended upon unconfirmed reports of African travelers. Du Chaillu himself tended to skepticism. He regarded gorillas as mostly herbivorous (correct, as we now know), despite numerous reports of frightening carnivory. Owen writes in 1865:

> Mr. Du Chaillu, however, states that he examined the stomachs of the Gorillas killed by himself and his hunters, and "never found traces there of aught but berries, pine-apple leaves, and other vegetable matter." The Gorilla is a huge feeder, as its vast paunch, protruding when it stands upright, shows.

Owen reports the lurid stories that du Chaillu heard from local people:

> Mr. Du Chaillu also adduces the testimony of the natives, that, when stealing through the gloomy shades of the tropical forest, they become sometimes aware of the proximity of one of these frightfully formidable Apes by the sudden disappearance of one of their companions, who is hoisted up into the tree, uttering, perhaps, a short choking cry. In a few minutes he falls to the ground a strangled corpse. The Gorilla, watching his opportunity, has let down his huge hind hand, seized the passing Negro by the neck with vise-like grip, has drawn him up to higher branches, and dropped him when his struggles had ceased.

But Owen also reports Du Chaillu's personal dubiety: "There is no doubt the Gorilla can do this, but that he does it I do not believe."

Du Chaillu's book of 1861 precipitated one of the greatest fracases in the contentious world of Victorian science. Many naturalists accused du Chaillu of fabricating tales; some suggested that he had never even visited the habitat of gorillas, but merely bought material shot by others. (Du Chaillu, for example, had written a dramatic account of shooting a large, enraged, and charging male—but the skin, shipped to Owen in London, bore no bullet holes in front.) Debate also raged on du Chaillu's claim for personal observation of a stunning habit that has defined our image of male gorillas ever since—pounding on the chest to display threat or anger. Owen reports the claim:

When so pursued as to be driven to stand at bay, the Gorilla, like the Bear, raises himself on his hind feet, with his powerful arms and hands free for the combat. In this predicament Mr. Chaillu affirms that the creature "offers defiance by beating his breast with his huge fists, till it resounds like a bass-drum."

In deference to both sides at once, and maintaining all options in the face of doubt, Owen then commented:

There is nothing in the structure of the Gorilla, save the size and depth of the chest, to suggest or accord with this peculiar action. Nor, were the dog as rare a beast, is there anything in its anatomy that would have suggested, to one who had never seen it alive, its occasional habit of running on three legs. In statements of this kind by a traveller, it is neither wise to discredit nor implicitly to believe; but one may acquiesce, and wait the report of succeeding observers whose attention has been directed to the original statement.

Most lurid legends turn out to be wrong, but du Chaillu was vindicated, and male gorillas do pound their chests, just like King Kong (but more in bluff than in prelude to battle). In fact, though du Chaillu did not emerge as a paragon of accuracy, he did fare well in the great debate, and he and his mentor Owen clearly won both a palm of victory and the right to thump.

As a curious footnote, du Chaillu's supporters also included a man so hostile to Owen, both for irreconcilable views and opposite personalities, that I doubt they ever again shared a common platform: England's most eloquent naturalist, Thomas Henry Huxley. Although Huxley found du Chaillu's book full of unintentional errors based on "imperfectly kept notes" and "a rather vivid imagination," he honored du Chaillu's courage in visiting dangerous and inaccessible places, and he found the explorer's accounts generally reliable. (Huxley later backed off in supporting du Chaillu, for he rankled at the mileage accruing to his enemy Owen, and just couldn't bear to act as an aide de camp.)

A Seahorse for All Races

Strange bedfellows do not only inhabit political hotels; science also spawns odd allegiances. Huxley and Owen could work together on du Chaillu's behalf because both needed information about gorillas to pursue their own disparate campaigns—based in large part on attempts to slaughter each other.

Owen published his major work on gorillas in the 1865 monograph given to Dickens, and in his Rede lecture of May 1859, ironically published in the same year as Darwin's *Origin of Species,* for Darwin's book would so recast the debate to Owen's disadvantage—"On the classification and geographical distribution of the Mammalia, to which is added an appendix 'On the Gorilla,' and 'On the extinction and transmutation of Species.'" Huxley featured gorillas in his finest and most influential publication, a landmark in the history of scientific prose: "Evidence as to Man's Place in Nature," a series of lectures originally given to working men in 1860 and 1862. (Admission supposedly required proof of blue-collar status, but legend proclaims that Karl Marx managed to sneak in!)

The grand Darwinian debate on evolution may have defined the broad subject of these volumes by Huxley and Owen, but their own excruciatingly bitter personal confrontation about apes and humans forms the controlling subtext for all these documents—and one cannot understand these works today without some background in "the great hippocampus debate." (Victorian scientists did pursue activities other than contentious argument, although the three altercations that act as pillars of this essay do form a totality—du Chaillu's gorilla rhubarb and the Huxley-Owen hippocampus rumpus as stalking-horses for the grandest of underlying issues, Darwin's brouhaha on the origin of species.)

I write this essay to memorialize the centenary of Huxley's death (1825–1895). The hippocampus debate has always been depicted as Huxley's greatest and absolutely unalloyed victory. I am also an unabashed Huxley fan, as illustrated by numerous essays in this series. As a fierce defender of evolution ("Darwin's bulldog" in the cliché), and the greatest prose stylist in the history of British science (though one might argue for a tie with D'Arcy

Thompson's *Growth and Form*), Huxley almost has to be my personal hero. Nonetheless, and following the antihagiographical bent of these essays, I choose the hippocampus debate to memorialize Huxley because I believe that the story has not been fairly told—and that, at one crucial point agonizingly relevant to current concerns, Owen developed an important and valued argument against a baleful implication of Huxley's generally admirable stance.

Many advantages accrue to the victors of any dispute, military or cerebral—and chronicling rights must rank among the greatest of perks. In short, the winners write history. How would we interpret the Trojan War if our main account had been written by Hector's bard; and how would future generations view the history of evolutionary theory if Duane Gish and Henry Morris (our most vociferous modern creationists) cornered the market for written descriptions?

Richard Owen (1804–1892) was the greatest anatomist and paleontologist of his age. His accomplishments were legion, both in range and excellence (including an early monograph on the chambered nautilus; the initial description of the oldest fossil bird, *Archaeopteryx;* a series of crucial papers on the moas, extinct giant ground birds of New Zealand; the first description of South American fossil mammals brought back by Darwin from his *Beagle* voyage; coining of the name "dinosaur," followed by volumes of accurate work on fossil reptiles of all ages). As a pillar of establishment science (and an intimate of Queen Victoria and nearly everyone else who mattered), Owen also wielded substantial power in the service of zoology, particularly in his long and successful campaign to establish a separate building for natural history within the British Museum. (Owen served as first director, and his edifice still stands, vigorously functioning in South Kensington as both a great monument of Victorian architecture and one of the most important scientific museums in the world.)

But Owen ran afoul of the ultimate victors in Victorian natural history— Darwin's circle. He was, to say the least, not a consistently nice man. He tended to be almost obsequiously genial and accommodating to those more

124

powerful than himself, but arrogant and dismissive toward juniors and underlings, the folks who eventually "grow up" and write later histories! He was not opposed to evolutionary ideas, despite later legends constructed by Darwinian chroniclers, though he strongly disliked Darwin's materialistic version of biological change.

The ever-genial Darwin wrote a most uncharacteristic assessment in his *Autobiography,* thus illustrating the offputting power of Owen's personality:

> I often saw Owen, whilst living in London, and admired him greatly, but was never able to understand his character and never became intimate with him. After the publication of the *Origin of Species* he became my bitter enemy, not owing to any quarrel between us, but as far as I could judge out of jealousy at its success. Poor Dear Falconer [a paleontological colleague], who was a charming man, had a very bad opinion of him, being convinced that he was not only ambitious, very envious and arrogant, but untruthful and dishonest. His power of hatred was certainly unsurpassed. When in former days I used to defend Owen, Falconer often said, "You will find him out some day," and so it has passed.

As a young man, eager to advance, Huxley bridled under Owen's power—and bided his time. Huxley, twenty years Owen's junior, often needed letters of recommendation from the purveyor of maximal patronage in natural history. Owen always complied, and with strong words of praise much to Huxley's advantage, but he always made Huxley wait, and treated him with condescension, if not contempt. Huxley recalled meeting Owen accidentally on the street after two unfulfilled requests for an urgently needed recommendation:

> I was in a considerable rage . . . I was going to walk past, but he stopped me, and in the blandest and most gracious manner said, "I have received your note. I shall grant it." The phrase and the implied condescension were quite "touching," so much so that if I

stopped for a moment longer I must knock him into the gutter. I therefore bowed and walked off.

Huxley and company eventually won rights to tell the official story, and they read Owen out of their triumphalist account, or (even worse) depicted him as a pompous fool, and an unwitting agent of their victory. But Owen has found modern defenders among historians out to debunk progressivist accounts of science as continuous advance fueled by saintly advocates of factual truth against infidels mired in social prejudice; Nicolaas A. Rupke's 1994 biography, *Richard Owen: Victorian Naturalist,* provides a splendid account in this corrective mode. Rupke quotes several genuine and warmly enthusiastic accounts of Owen, from many sensible and admirable people (including Charles Dickens). Let us at least acknowledge that Owen was an enormously complex, brilliant, and fascinating man—and that the history of biology in Victorian Britain cannot be told without granting him a long chapter.

The great hippocampus debate began two years before Darwin published the *Origin of Species,* and did not invoke natural selection as a central subject. But this most famous Victorian scientific wrangle did engage the primary, and perpetually gut-wrenching subject that Darwinism placed into such sharp focus: the uniqueness of humans among other animals. Are we just improved apes, or something entirely different from all other creatures? Huxley advocated continuity with gorillas; Owen defended sharp separation. The turf of battle, by Owen's choice and initial proposal, centered upon three structures in the brain. Owen claimed that only humans possessed these features, thus defining our absolute separation from the brute creation. Huxley proved that apes possessed versions of all three structures—sometimes as prominently expressed as in humans—and that Owen's marks of separation therefore affirmed our evolutionary unity with other primates.

For the first difference, Owen claimed that only humans possessed a "posterior lobe" of the cerebrum—a backward extension of this traditional location for "higher" mental functions—to cover the cerebellum, or conventional region for control of motor activity. (The accompanying figure from Owen's 1859 lecture will clarify his claim. Note that the chimp cerebrum [letter A] stops in front of the underlying cerebellum [letter C], but extends to cover the brain's entire upper surface in humans. By modern neurological evidence, the traditional attributions of function are as wrong as Owen's claims for morphological differences, but I cite the older views to situate the debate in its own time. Obviously, if the "higher" cerebrum covered the "lower" cerebellum only in humans, we might measure our mental superiority thereby.)

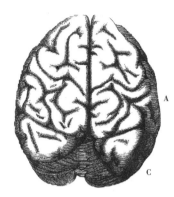

Owen's drawing of a chimpanzee's brain, showing that the cerebrum does not extend back far enough to cover the cerebellum (labeled C).

Second, Owen stated that only humans possess a posterior cornu in the lateral ventricle. (To explicate this mouthful, ventricles are spaces within the brain, continuous with the central cavity of the spinal cord and formed as the developing brain undergoes complex bending and folding in embryology. *Cornu* is Latin for "horn"—so the posterior cornu is a horn-shaped rear end to a cavity in the brain.)

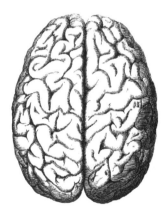

In humans, the larger cerebrum completely covers the cerebellum.

Third and last, Owen claimed that only humans developed a "hippocampus minor"—a ridge on the

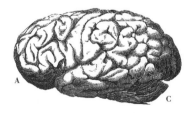

bottom of the same posterior horn of the lateral ventricle, and produced by a deep inward penetration of an adjacent part of the brain called the calcarine fissure. This "hippocampus minor" is not the same structure as the "hippocampus" itself, an important region of the brain's old interior, recently identified in a series of elegant experiments as a site for the initial recording of short-term memories, which are then, somehow, transferred to the neocortex for long-term storage. In modern terminology, "hippocampus minor" has been dropped in favor of the earlier name, *calcar avis,* or "cock's spur," in reference to visual similarity with the weapon on a rooster's leg that potentiated the "sport" of cockfighting. The name *hippocampus* was originally coined in the sixteenth century by Arantius, a student of Vesalius, because the structure reminded him of a seahorse—*Hippocampus* in Latin, and the formal name later chosen by Linnaeus to designate the major genus of seahorses.

I will not rehearse the details of the hippocampus debate here, for no story in Victorian natural history has been told so often, and the basic facts are not in dispute. (My friend and colleague Charles G. Gross, professor of psychology at Princeton, has recently published a particularly clear and accessible article, notable for its focus on anatomical details of the brain—for Gross is a celebrated neurologist, not primarily a historian: "Hippocampus Minor and Man's Place in Nature: A Case Study in the Social Construction of Neuroanatomy," published in *Hippocampus.* Yes, in our world of specialization, each region of the brain has a journal devoted to its study!)

The debate arose in the late 1850s and lasted with full force into the 1860s, sputtering out by the time Owen wrote his gorilla monograph in 1865. Owen and Huxley duked it out both in writing and in public appearances, notably at the same 1860 British Association meeting where, according to legend and unsupported by fact, Huxley also destroyed Bishop "Soapy Sam" Wilberforce in an initial altercation over Darwinism (the exchange took place, but without a clear victor). The debate spilled vigorously and copiously into general culture, as the public and press delighted in watching two of Britain's greatest scientists acrimoniously debating such an important issue (the status of humans in nature) by wrangling about parts of the brain unknown to all and endowed with such wonderfully amusing names as "hippocampus minor." Charles Kingsley, featuring the hippocampus debate in his 1863 children's classic *The Water Babies,* emphasized the humor implicit in the conjunction of arcane anatomical mumbo-jumbo with a theme of such conceptual and emotional importance. Kingsley writes of Professor Ptthmllnsprts (Put-them-all-in-spirits), his parody of Huxley:

> He held very strange theories about a good many things. He had even got up once at the British Association, and declared that apes had hippopotamus majors in their brains just as men have. Which was a shocking thing to say; for, if it were so, what would become of the faith, hope, and charity of immortal millions? You may think that there are other more important differences between you and an ape, such as being able to speak, and make machines, and know right from wrong, and say your prayers, and other little matters of that kind; but that is a child's fancy, my dear. Nothing is to be depended upon but the great hippopotamus test.

On May 18, 1861, *Punch* published a quite accurate account of the issue in doggerel verse, beginning:

> *Then Huxley and Owen*
> *With rivalry glowing,*
> *With pen and ink rush to the scratch:*
> *'Tis Brain versus Brain,*
> *Till one of them's slain;*
> *By Jove! it will be a good match!*

And indeed it was! Private letters give a good account of the animosity. Huxley wrote that he would "nail that mendacious humbug . . . like a kite to the barn door." Owen described one of their public altercations to a friend: "Prof. Huxley disgraced the discussions by which scientific differences of opinion are rectified by imputing falsehood on a matter in which he differed from me. Until he retracts this imputation as publicly as he made it I must continue to believe that, in making it, he was merely imputing his own base and mendacious nature."

The "official" account of the debate can be summarized in a paragraph: Huxley approached the controversy like a military general, out to upstage an older enemy. He organized several colleagues to dissect the brains of various apes and monkeys in search of the structures that Owen had pronounced unique to humans. Huxley himself studied the brain of the South American spider monkey, a "lower" primate on the traditional scale. They also scoured published literature, searching both for Owen's distortions or selective quotations, and for prior proof of the three structures in nonhuman primates. In short, they found abundant evidence for all three structures in various primates. Owen, according to legend, eventually shut up and licked his wounds.

Owen's tactics clearly led him to disaster. (Both Owen and Huxley came from lower middle-class backgrounds, but Owen had ingratiated himself into upper circles, to his enormous practical benefit, receiving a rent-free residence directly from Queen Victoria and, beginning in the early 1840s, an annual civil list pension, all achieved while Huxley struggled financially and grew bitterly jealous. Never doubt the centrality of social class for understanding Victorian life.) As a *nouveau arrivé* in the upper classes, Owen felt that he had to obey perceived norms for imperious disregard of upstarts. He fought for his positions, and against Huxley, but not with the same vigor, and with none of Huxley's overt organization. Owen largely tut-tutted among his noble friends, and lost precious ground.

Owen's tactics evidently failed, but were his arguments so bad? I will not attempt to deny the usual reading—that Huxley and his phalanx found all three structures in the brains of nonhuman primates, therefore disproving Owen's criteria of uniqueness. But was Owen really so stupid, and so defeated? I don't think so. I would like to raise two points for balance, and partly in Owen's defense—the first well discussed by Rupke and others, the second (so far as I know) not previously covered in the literature.

First, how did Owen respond to Huxley and company's discovery of the three structures in apes and monkeys? Did he just deny their findings, or suffer in silence? In fact, Owen made a potentially fair reply. He had been stating for years (though he had conveniently omitted the claim in several publications at the height of the hippocampus debate) that virtually every feature of humans has a homologous expression in closely related chimps and gorillas. (Owen had coined the term *homology* in the late 1840s to iden-tify features of identical anatomical origin in different creatures, whatever the degree of functional divergence—wings of bats with front legs of horses, for example. We now attribute homology to evolutionary descent from a common ancestor.) But admission of homology does not require application of the same name to the relevant feature in two organisms, for functional divergence might legitimately permit a different term. For example, calling the bat's forearm a wing doesn't obligate us to state that all mammals have wings because all carry homologs of the bat's forearm bones.

Owen invoked a sneaky version of this purely terminological point to worm out of his defeat—but we must at least credit the technical validity of his claim. In the 1865 gorilla monograph, he admits that the three structures do exist in apes, but in such rudimentary state, and in such different form from their expression in humans, that all must receive a different name— just as horses don't have wings. Thus, we can still say that apes don't have a posterior lobe, a posterior cornu, and a hippocampus minor—even though they possess homologs of all these human structures!

Owen begins by allowing pervasive homology: "In the Gorilla . . . the homologue of every organ and of almost every named part in Human anatomy is present." He then discusses how a gorilla brain might be topo-logically transformed to a human brain, almost nonchalantly admitting at the end that the three structures already exist in gorillas!

> . . . to expand the cerebrum in all directions, and especially backward beyond the cerebellum, so as to define a "posterior" or "post cerebellar" lobe: to extend the chief cerebral cavity, or "lateral ventricle" . . . backward . . . into a "posterior horn" . . . with prominences corresponding with . . . the "hippocampus minor"; the beginnings or incipient homologues of which cavity and part are alone present in the highest Apes.

I could accept Owen's redefinition as clever and honorable, but for two points. First, he implies that he doesn't have to give the same name to the three features in apes because they are so poorly developed. But Huxley and colleagues had shown that some apes develop these features to equal strength with their expression in humans. Second, if Owen had taken this position all along, then we could blame Huxley for unsubtlety. But, in fact, Owen didn't begin by admitting rudimentary homologs of the three features in apes, and merely claiming that strong human development required a separate name. He really did deny that the structures existed *at all* in apes. Owen wrote in his 1859 lecture:

> Posterior development [of the cerebrum] is so marked that anthropotomists have assigned to that part the character and name of a "third lobe"; it is peculiar and common to the genus *Homo:* equally peculiar is the "posterior horn of the lateral ventricle" and the "hippocampus minor," which characterize the hind lobe of each hemisphere. Peculiar mental powers are associated with this highest form of brain . . . I am led to regard the genus *Homo* as not merely a representative of a distinct order, but of a distinct subclass, of the Mammalia.

I do think, particularly in this context of focal conclusion, that "peculiar" means "unique"—and that Owen did shift his ground by verbal ploy, thus illustrating the valid part of Huxley's mistrust and anger.

But the second point, previously overlooked in our large literature, does corroborate an important part of Owen's argument—one with wrenching implications in our current reality. Huxley was clearly right in demonstrating the three structures in other apes, but his central presentation of the argument for evolutionary continuity between apes and humans, published most prominently in *Man's Place in Nature* (1863), rests upon two false arguments (one as sneaky as Owen's later attempt to cover his ass), and both well refuted by Owen in his gorilla monograph of 1865.

Man's Place in Nature presents the strongest defensible version of smooth evolutionary transition between apes and humans. Huxley admits the undeniable gap in brain size, with smaller-bodied humans carrying a brain three times as large as that of much weightier gorillas, but correctly identifies this disparity as a gulf in quantity alone, for all parts of the brain are homologous in apes and humans. Huxley then, and with equal justice, argues that a different *quantity* of brain need not account for a true gulf in *quality* of mental operations, for such a claim would confuse correlation with causality. Perhaps, Huxley states, human cognitive superiority resides in some unidentified difference of cellular or microarchitectural function, and not in disparity of bulk alone.

Huxley then presents his two linked arguments. First, he hammers home by brute force, listing feature after feature, his key claim for continuity: the gap between the lowest ape and the chimp or gorilla is far greater than the corresponding gap between these "highest" apes and humans—so we are just one small step farther along in the sequence of apes:

> Whatever part of the animal fabric—whatever series of muscles, whatever viscera might be selected for comparison—the result would be the same—the lower Apes and the Gorilla would differ more than the Gorilla and the Man . . . The structural differences between Man and the highest Apes are of less value than those between the highest and the lower Apes.

But Huxley's argument seems unfair and even a tad self-serving within the context that both he and Owen shared: a belief that groups of related organisms should be ranked on a scale from lower to higher. (Our current denial of this scheme is, of course, irrelevant to our analysis of the logical structure of Huxley and Owen's arguments.) Owen properly refuted Huxley by pointing out that he had made a false comparison of disparate things—apples and oranges in the current cliché. The gap between gorilla and human amounts to one step, but the separation of "lower" primates from chimps and gorillas encompasses scores of omitted intermediates. If I were trying to minimize the gap between step 50 and 51 in a series by arguing that the separation between step 1 and step 50 is even greater, you would properly laugh me to scorn and say, "Don't load the dice inevitably in your favor; compare the right things. Tell me about the gap between 49 and 50 versus 50 and 51. A step must be compared with a step, not with the entire series!" If the gorilla-human step exceeds the distance between any two adjacent primates, then I may regard humans as something special. Owen writes:

> Passing . . . to a comparison of the Gorilla's brain with that of other Quadrumana [apes], we discern the importance and significance of the much greater difference between the highest Ape and lowest Man, than exists between any two genera of Quadrumana in this respect . . . From [gorillas] to Lemurs [the "lowest" primate in Owen's scheme] the difference of cerebral development shown in any step of the descensive series is insignificant compared with the great and abrupt rise in cerebral development met with in comparing the brain of the Gorilla with that of the lowest of the Human races.

I think that Huxley sensed the weakness of his argument, for he introduced a second supposed clincher in conjunction and support: the gap between gorilla and average human may be large, but if we order all human variation in a hierarchical ranking of races, from the "lowest" Negro to the "highest" Caucasian, then the gap closes, for the step from gorilla to lowest human becomes less than the space between lowest and highest *Homo sapiens*. (Please understand that I am using Huxley's own terms and quoting the conventional wisdom of his day within the restricted community of scientists—that is, white males of privilege.) Huxley wrote:

> The difference in weight of brain between the highest and the lowest men is far greater, both relatively and absolutely, than that between the lowest man and the highest ape . . . Thus, even in the important matter of cranial capacity, Men differ more widely from one another than they do from the Apes.

I do not think that Huxley, a racial liberal by the standards of his time, advanced this argument with intent to impugn entire groups of human beings. Rather, he was trying to plug a hole in his central argument for evolutionary continuity by finding some way to fill the embarrassingly large space in cranial capacity between gorilla and average human.

This complex world of ours, this vale of tears, lies awash in irony. Just as bad things happen to good people, decent folks also advance logically fallacious and morally dubious claims in support of good arguments. Huxley stood on the side of the angels: he tried to advance the cause of human evolution by documenting continuity with our closest animal relatives. He closed his case with a magnificent prose flourish in describing the great range of design within the primate order, from the lowest lemur to our exalted selves:

> Perhaps no order of mammals presents us with so extraordinary a series of gradations as this—leading us insensibly from the crown and summit of the animal creation down to creatures, from which there is but a step, as it seems, to the lowest, smallest, and least intelligent of the placental Mammalia. It is as if nature herself had foreseen the arrogance of man, and with Roman severity had provided that his intellect, by its very triumphs, should call into prominence the slaves, admonishing the conqueror that he is but dust.

Nonetheless, and whatever his broader good intent, Huxley did advance in support a harsh, uncompromising, and undeniably racist argument that arranged all humans in a line of advancing worth, and explicitly identified African blacks as midway at best between gorillas and European whites. Huxley's error arose from a deep fallacy in his evolutionary reasoning—the progressivist equation of evolution with linear advance. He assumed that evolution must proceed in a series of rising steps, and he felt that he couldn't defend human evolution unless he could demonstrate such a linear order among modern people. In this assumption, he committed an even deeper error based on a classically false premise of reasoning: belief in continuity of cause, with failure to recognize that superficially similar phenomena at different scales may have disparate causes.

Yes, humans differ from apes and, yes, humans vary among themselves. But these facts don't imply that variation among modern humans acts as a microcosm for larger differences between humans and other species—though Huxley assumed such continuity when he ran human racial variation along the same scale as differences among primate species. Human races are not surrogates for intermediate steps between ancestral apes and modern people; human races represent an entirely different scale of contemporary variation within a single biological species. We have no reason either to rank variation within a species along any line of worth, or to regard such contemporary diversity as particularly related to modes of our evolutionary derivation. Of course, evolution does predict that the gap between ancestral apes and modern humans must be bridgeable, but the transitional forms are extinct species of the fossil record, not modern races. Moreover, since modern races are so young (as we now know), our differences are effectively inconsequential in evolutionary terms. No human race is, in toto, more apelike than any other. We are all recently derived varieties of the common human stock, *Homo sapiens.*

Poor, maligned, politically conservative, intellectually antediluvian Richard Owen. He took one look at Huxley's racist argument, and nailed him—like a kite to the barn door, and exactly for the right reasons. I know that Owen did not refute Huxley in the service of racial egalitarianism. I know that Owen shared all Huxley's prejudices about racial ranking and the existence of higher and lower forms of human life. Owen's text is sprinkled with the conventional language of a shared racist perspective. In 1859 he wrote that the chimpanzee lies "nearer than any other known mammalian animal to the human species, particularly to the lower, or Negro forms." And later in the same work: "In the low, uneducated, uncivilized races, the brain is rather smaller than in the higher, more civilized and educated races." In the 1865 gorilla monograph, he rolls all common prejudices into one line by stating that male skulls must be treated as standards, with both females and lower races (identified, conventionally, as Ethiopian, or African black, and Papuan, or Melanesian black) as inferior: "If the naturalist . . . were to abandon his proper guide, viz. the average condition of the brain in the male sex, and to take the brain of a female of the lowest Papuan or Ethiopian variety . . ."

I also know that Owen refuted Huxley's racist argument in order to defend human uniqueness against a claim for continuity, and not for any social or political motive that we might honor today. Nonetheless, intentions and consequences must be separated (and much of the fascinating complexity and moral ambiguity in our lives arises from the sharp disparity so often encountered between our goals and the opposite, yet unavoidable, side consequences of actions taken in the service of these goals—oppose hunting on principle, and too many deer may eat your flower gardens). Thus, I applaud the consequence of Owen's argument, whatever his intent.

Moreover, in this particular case, Owen's refutation of a racist argument did not arise accidentally from a claim advanced entirely for other reasons. I honor Owen even more because he knew exactly what he was doing—as he directly quoted the few egalitarian sources of his time, and explicitly advanced his claim in the service of racial melioration (though not equality—an option that, sadly, did not exist in Owen's intellectual framework).

By identifying human racial variation as both small in extent and fully encompassed within an indivisible species—in other words, as something different from the gaps between species—Owen refuted Huxley's second crucial point, that the gap from highest ape to lowest man did not exceed the space between lowest and highest men. In the key passage, he writes:

> The extent of differences in the proportion of the cerebrum . . . in the different varieties of mankind is small, and with such slight gradational steps as to mark the unity of the human family in a striking manner.

But the most important sentence occurs two pages earlier:

> Although in most cases the Negro's brain is less than that of the European, I have observed individuals of the Negro race in whom the brain was as large as the average one of Caucasians; and I concur with the great physiologist of Heidelberg, who has recorded similar observations, in connecting with such cerebral development the fact that there has been no province of intellectual activity in which individuals of the pure Negro race have not distinguished themselves.

Owen then appends an interesting footnote:

> The University of Oxford worthily conferred, in June 1864, the degree of Doctor of Divinity on Bishop Crowther, a member of pure West African Negro race, who was taken from his native land as a slave, and recaptured in the middle passage. I record with pleasure the instruction I have received in conversation with this sagacious and accomplished gentleman.

(Samuel Adjai Crowther, 1812–1891, was captured from a slaving ship by a British man-of-war in 1822 and returned as a free man to Sierra Leone. Baptized in 1825, he attended mission schools in Africa, and then traveled to England, where he was ordained in 1842 and consecrated in 1864. He then served as bishop of Niger territory, where he translated the Bible into Yoruba.)

Owen's passage surely reeks with paternalism by irrelevant modern standards, but we should honor his decency at a time when some colleagues wouldn't ever deign to socialize with a black man. Owen's most revealing words, however, refer to "the great physiologist of Heidelberg"—for here we do grasp his unconventional allegiances. Friedrich Tiedemann, professor of anatomy at Heidelberg, was the only genuine egalitarian among early-nineteenth-century European scientists of eminence. He measured skulls of all races and wrote several treatises on the putative intellectual equality of all people. He submitted a major article in English to the *Philosophical Transactions of the Royal Society* in 1836, the document quoted by Owen. If Owen explicitly cited Tiedemann, we can be confident that he chose to refute Huxley's argument on race, at least in part, by defending the high intellectual achievement and capacity of all human groups.

From 1859 until his death in 1870, Charles Dickens published a weekly miscellany of literature and current events entitled *All the Year Round*. He did not write each piece himself, but he exercised such a strong editorial hand that the *Encyclopaedia Britannica* remarks: "He took responsibility for all the opinions expressed (for authors were anonymous) and selected and amended contributions accordingly. Thus comments on topical events may generally be taken as expressing his opinions, whether or not he wrote them." Dickens published his major commentary on Darwin in the July 7, 1860, issue of *All the Year Round*. The closing paragraph reads:

> Timid persons, who purposely cultivate a certain inertia of mind, and who love to cling to their preconceived ideas, fearing to look at such a mighty subject from an unauthorized and unwonted point of view, may be reassured by the reflection that, for theories, as for organized beings, there is also a Natural Selection and a Struggle for Life. The world has seen all sorts of theories rise, have their day, and fall into neglect.

Owen's theory fell and died; Huxley's views prevailed, both by virtue of essential truth and possession of the right to tell history. But amalgamations usually forge our best solutions in a complex world—and I wish we had preserved Owen's correct and principled argument on race in proper integration with Huxley's evolutionary perspective. Such a conjunction, if incorporated into political and social policy as well, could have spared human history from most major horrors of the past century. We must still struggle to craft the conjunction, a tale of two worlds—for, in so doing, we might convert the "worst of times" to "the best of times," an "age of foolishness" into an "age of wisdom," and the "season of Darkness" into a "season of Light."

7

MR. SOPHIA'S PONY

IF A STOLEN PURSE COUNTS ONLY AS TRASH COMPARED WITH A GOOD NAME lost, how shall we judge the happily expiring custom of addressing a married woman by her husband's name? Perhaps I was an incipient feminist from my cradle, but I do remember wondering, at a very early age, why my mother, Eleanor, often received letters addressed to a Mrs. Leonard Gould.

Among several possible redresses, the game of turning tables in favorable circumstances surely has appeal. Samuel Gridley Howe (1801–1876) did good work as an educator of the blind, but I once took great pleasure in identifying him as Mr. Julia Ward Howe to acknowledge his more famous wife, author of "The Battle Hymn of the Republic."

We do not often encounter such an opportunity among married scientists, given the virtual exclusion of women before our current generation. Mme. Curie stands among the greatest scientists of all time, but her husband, Pierre, was also no slouch, and must therefore remain Pierre, not Mr. Marie. But I do know one scientific couple subject to this strategy of inversion for pairing an eminent wife with an obscure husband—and I do feel quite protective, for Mr. Sophia Kovalevsky was a paleontologist, and a damned fine (if forgotten) scientist in his own right.

The *Dictionary of Scientific Biography* begins its entry on Sophia Kovalevsky (1850–1891) by calling her "the greatest woman mathematician prior to the twentieth century." She studied abroad, for women could not obtain degrees from Russian universities. In Berlin, she received four years of private tutoring from professors, for women could not attend university lectures. In 1874, she earned her doctorate in absentia from the University of Göttingen in Germany. Despite the acknowledged excellence of her research, and solely for reasons of her gender, Kovalevsky could not obtain an academic position anywhere in Europe. She therefore returned to Russia—to a life of odd jobs, failed business ventures, and stolen hours for mathematical study. In 1883, following the death of Mr. Sophia—we shall come to him shortly—she again tried to obtain an academic post, this time successfully. She enjoyed several years of productive work as a professor of mathematics at the University of Stockholm, but died of influenza and subsequent pneumonia at age forty-one, at the height of her accomplishments and fame.

Sophia Kovalevsky published only ten papers on mathematical subjects during her brief life (she also attempted, less successfully, a simultaneous career in literature, writing several novels, a play, and a critical commentary on George Eliot, whom she had met on a trip to England). But these substantial works on diverse problems in mathematics brought her much renown. She studied the propagation of light in a crystalline medium, the rings of Saturn, and the rotation of rigid bodies around fixed points; she wrote several papers on technical matters (that I do not pretend to understand) in integral calculus. In 1888, she won the Boudin Prize of the French Academy of Science for her memoir on the rotation of rigid bodies (which generalized the work of her French predecessors Poisson and Lagrange). The judges were so impressed by her research that they raised the award from three thousand to five thousand francs to express their gratitude.

Mr. Sophia's Pony

Vladimir Onufrievich Kovalevsky (1842–1883)—Mr. Sophia—entered his wife's life in a most unromantic, but eminently practical, manner integral to this tale of Sophia's career. Single women of intellectual bent languished in Catch-22 in mid-nineteenth-century Russia. They could not study at Russian universities, but they could not travel abroad as independent persons, either. To escape this bind, freethinking women often arranged sham marriages with men of similar persuasions. The technically married couple could then travel abroad for foreign study. Sophia wed Vladimir for emancipation and the right to travel. The newlyweds then left for Germany, to live in different apartments and study in different cities.

Sophia and Vladimir belonged to the culture of freethinking Russians who, in pre-revolutionary times, gave their name to one of the few English words with a Russian etymology—the intelligentsia. The men and women of the intelligentsia tended to radicalism in politics, Bohemianism in lifestyle (in stark contrast to the proclaimed asceticism of later Bolshevism), and—in a striking difference from American and Western European versions of the same phenomenon—fascination for science and confidence in its power to transform the world for good. These men and women sought scientists for their idols, rather than the literary or philosophical intellectuals who fronted for similar movements in other lands. Darwin, in particular, became their icon—and for this reason (and as a historical curiosity rarely acknowledged or appreciated), most Russian intellectuals were strict Darwinians while, to Darwin's own frustration, other European scientists, although convinced of evolution's truth by the *Origin of Species,* tended to reject Darwin's favored mechanism of natural selection.

As a literary prototype of this movement, we must nominate the hero Bazarov of Turgenev's *Fathers and Sons,* published in 1862, just three years after Darwin's *Origin of Species.* This revolutionary nihilist denies all laws except those of the natural sciences. When not engaged in political schemes, he dissects frogs to build his knowledge and center his life. Vladimir and Sophia Kovalevsky were not nearly so colorful, or extreme in their sentiments and actions. But their lives surely included sufficient adventure for a

Hollywood biography. I was particularly struck by the story of Vladimir's skullduggery in fostering the escape from France, following the fall of the Paris Commune in 1871, of the imprisoned and politically radical lover of Sophia's sister.

Sophia and Vladimir began their marriage as a sham, but as the world turns, birds fly, and bees buzz, the best-laid plans of mice and men often depart from original intentions. When Sophia could not find employment in Europe, and Vladimir wished to go home (where he could work as a pale-ontologist), they returned to Russia together. They had always been fond of each other, and when Vladimir showed special tenderness to Sophia follow-ing the death of her beloved father, they did consummate their marriage and eventually had a daughter, who later studied medicine, worked as a translator, and became quite a heroine herself within a very different Soviet system.

Their life together in Russia produced little but tension, much of their own making. Vladimir had some family money, and Sophia obtained a good sum after her father's death. As neither found remunerative work within science (and since they had chosen a lifestyle far above their means), they invested their cash in a variety of ill-considered business schemes, mostly in real estate and public baths—and quickly became flat broke. Vladimir then had a stroke of good fortune that eventually became his undoing. He obtained a position—at decent compensation—as spokesman for a firm that manufactured naphtha from petroleum. The Ragozin brothers, owners of the company, wanted the prestige of Vladimir's academic degrees, and his recognized verbal skills (arising, in part, from his efforts as a formidable soapbox orator in his political past), to lure customers and investors.

Vladimir spent most of his time on business trips to European cities. Sophia, though happy for the cash that gave her some leisure for mathe-matical work, became increasingly frustrated at his absences and preoccu-pations—and a serious rift developed. Finally, early in 1881, Sophia boiled over and left for Berlin to pursue her academic dreams. She explained in a letter to Vladimir:

You write truly that no woman has created anything important, but it is just because of this that it is essential for me, while I still have energy and tolerable material circumstances, to position myself so that I may show whether I can achieve anything or whether I lack brains.

(I have become fascinated with Vladimir and Sophia and have read everything I could get my hands on for this essay—including a veiled biography by Sophia's sister, an even more hagiographical set of Soviet documents, and a fine modern biography, the source of this quote and much else, by Don H. Kennedy: *Little Sparrow: A Portrait of Sophia Kovalevsky*.)

After many heartrending letters, and a few meetings for attempted reconciliation, Sophia decided to remain abroad, and the couple entrusted their daughter to the care of Vladimir's more famous (and solvent) brother Alexander (the celebrated embryologist who discovered the relationship between vertebrates and the apparently "lowly" marine tunicates).

And the predictable tragedy finally unfolded. Vladimir had been mentally ill for years, and his periods of depression lengthened and deepened. The naphtha firm failed, and the Ragozin brothers, charged with numerous shady dealings, faced judicial proceedings. Vladimir, fearing his own disgrace and prosecution (though he was apparently innocent and not under official suspicion), committed suicide on April 15, 1883, by putting a bag over his head and inhaling chloroform. He had earlier written (but not mailed) a letter to his brother that functioned as a suicide note:

> I am afraid that I shall grieve you very, very much, but from all the clouds that have gathered from all sides over me, this was the only thing left for me to do. Everything for which I was preparing has been broken up by this, and life is growing terribly difficult . . . Write Sophia that my constant thought was about her, and how very wrong I was before her, and how I spoiled her life which, except for me, would have been bright and happy.

On learning the news, Sophia was devastated by a complex mixture of grief and guilt. She withdrew to her room and would neither admit anyone nor eat anything. On the fifth day, she lost consciousness. She was then force-fed by her physician and put into bed. Several days later, she sat up, asked for a pencil and paper, and began to work on a problem in mathematics.

Vladimir's paleontological career was brief and limited, both in quantity and apparent range of material. He worked and studied abroad from 1869 (the year of his marriage) to 1874, attending lectures in several German universities, studying fossil vertebrates at museums in Germany, France, Holland, and Great Britain, and collecting fossils in France and Italy. He wrote six papers in three languages, none his own (a few Russian translations appeared later). All six, published between 1873 and 1877, treated the anatomy and evolution of the two great groups of large, hoofed, herbivorous mammals—the perissodactyls, or odd-toed ungulates (represented today only by the few living species of horses, rhinoceroses, and tapirs), and the artiodactyls, or even-toed ungulates (the greatest success story among large mammals, including the highly diverse cattle, deer, antelopes, sheep and goats, pigs, giraffes, camels, and hippos).

In the late 1970s, I edited an ill-fated thirty-volume collection of facsimile reprints in the history of paleontology. (The volumes were lovely, but the press went belly-up—for other reasons, I trust!—soon after the collection appeared. I assume that most printed copies ended up in the shredder.) I decided to collect all of Kovalevsky's papers together in one volume—his articles in German from the journal *Palaeontographica;* his English monograph, submitted by T. H. Huxley for publication in the *Philosophical Transactions of the Royal Society of London;* and his famous French treatise on the evolution of horses, published back home in the *Mémoires de l'Academie Impériale des Sciences de St.-Pétersbourg* (but not in Russian, so that wide readership would be possible. Many nations today—particularly Japan—publish major scientific journals in English for the same reason). The six papers made a substantial, but not particularly hefty, volume.

146

Mr. Sophia's Pony

Such limited output rarely builds preeminence in a field like paleontology, so identified (however unfairly) as a profession devoted to the detailed description of minutiae. Yet, although Vladimir Kovalevsky may be virtually unknown to the larger intellectual world (except as the husband of a famous mathematician, and brother of a celebrated embryologist), he is treasured within my small fraternity as an important innovator and a particularly careful craftsman. His few published papers created a reputation well beyond their literal heft. For fifteen years, I have stared fondly at my modest volume of his totality.

Kovalevsky has always won warm accolades from aficionados. Darwin greatly admired his work, and singled out a monograph on horses for special praise in a letter that can only make scholars yearn for more in the same (literal) style. (Darwin had the world's most abominable handwriting, a serious impediment for all historians of science. Some of the most important passages in his writing, and therefore in the history of Western thought, have yet to be deciphered to everyone's satisfaction. But he wrote to Kovalevsky in a wonderfully clear hand—no doubt laboriously and in deference to a man who usually worked in Cyrillic and might have difficulty with English penmanship. Why didn't Darwin realize that Englishmen might also stumble over his usual scrawl?)

Darwin had good reason to cultivate Kovalevsky's favor. Before marrying Sophia, Vladimir had worked as a translator and publisher of scientific books. He translated at least three of Darwin's most important works into Russian—*The Variation of Animals and Plants Under Domestication* (1868), the *Descent of Man* (1871), and the *Expression of the Emotions in Man and Animals* (1872). Vladimir worked so feverishly on the 1868 book (Darwin's longest) that the Russian edition actually appeared before the "original" English version, thereby marking the premiere of this important work. In another tale from their eventful lives, Vladimir and Sophia safely carried the proofs of the *Descent of Man* through the Prussian lines and into besieged Paris during the Franco-Prussian war of 1870.

Kovalevsky's reputation has always remained firm in the small fraternity of vertebrate paleontologists. In the first years of this century, Henry Fairfield Osborn, America's leader in the field, marked Kovalevsky's work as "the first attempt at an arrangement of a great group of mammals upon the basis of the descent theory." He then added:

> If a student asks me how to study paleontology, I can do no better than direct him to the *Versuch einer natürlichen Classification der fossilen Hufthiere* [*An Attempt at a Natural Classification of Fossil Hoofed Mammals*—Kovalevsky's most important German publication] . . . This work is a model union of the detailed study of form and function with theory.

The Belgian paleontologist Louis Dollo, Europe's leader in the field, praised Kovalevsky as "the first to study systematically the great problems of paleontology on the basis of evolution . . . No paleontologist had ever joined such an intimate knowledge of details with such an amplitude of concepts." In his major work (*La paléontologie éthologique*, 1909), Dollo proposed a heroic epitome by depicting the history of paleontology as a forward march from foolishness to illumination in three progressive stages, each marked by a prototype: the *"époque fabuleuse, ou empirique,"* centered upon the fanciful early-eighteenth-century work of the Swiss savant J. J. Scheuchzer; the *"époque morphologique, ou rationelle,"* marked by the great Georges Cuvier in the early nineteenth century; and the culminating *"époque transformiste* [evolutionary], *ou définitive,"* symbolized by the brilliant work of Kovalevsky.

We may identify the source of Kovalevsky's fame, and his enduring place in the history of the natural sciences, in two summary statements.

1. Kovalevsky made the first substantial application of evolutionary theory— specifically, Darwin's version based on natural selection—to lineages of fossil organisms. (Others had published evolutionary interpretations based on vague and confused views about mechanisms of change, but Kovalevsky rig-

orously applied Darwin's theory of natural selection, primarily by seeking correlations of altered anatomy with changing external environment, and then developing functional, or adaptive, interpretations of the evolutionary transformations.) Moreover, Kovalevsky was a consistent Darwinian at a time when the great majority of scientists, although persuaded of evolution's truth by Darwin's arguments, rejected natural selection as an important mechanism of change. (In fairness, Kovalevsky's commitment need not record any superior insight or observation based on fossils, but arose largely from Darwin's exalted status among the Russian intelligentsia, as discussed previously in this essay.)

In his English monograph of 1874, Kovalevsky wrote:

> The wide acceptance by thinking naturalists of Darwin's theory has given a new life to paleontological research; the investigation of fossil forms has been elevated from a merely inquisitive study of what were deemed to be arbitrary acts of creation to a deep scientific investigation of forms allied naturally and in direct connection with those now peopling the globe.

As a good example of his focus on adaptation to changing environments as the motor of evolution, Kovalevsky argues, later in the same monograph, that horses evolved their strong single toe for life on hard, dry plains (a new environment following the evolution of grasses in the Miocene epoch), while hoofed mammals living in soft and swampy ground required a broad foot and therefore retained several toes:

> If the lateral digits are still retained in the Suidae [pigs] it is chiefly owing to the fact that the Hogs live generally in marshy places and on muddy river-banks, where a broad foot is of great importance for not allowing them to sink deeply into the mud. But if, by some geological change, their habitat should be transformed into dry grassy plains, there can be no reasonable doubt that they would as

readily lose their lateral digits as the Paleotheroids [horse ancestors, in Kovalevsky's view] have lost theirs . . . in becoming transformed into the monodactyle [one-toed] Horse.

2. Kovalevsky documented the most famous evolutionary story of all—the transformation of a small, many-toed ancestor with low-crowned teeth into the large-bodied, single-toed, long-toothed modern horse. Moreover, Kovalevsky identified, correctly we still think, the primary adaptive basis of this transformation—an environmental shift from browsing leaves in woodlands and marshes, to grazing of grasses and running on the open plains. Kovalevsky tied this transition to the Miocene evolution of grasses, and the subsequent development of extensive plains and savannahs as a new habitat ripe for exploitation. He explained reduction of toes as an adaptation for running on hard ground, and the evolution of high-crowned teeth as a necessary response to new diets of tough grasses with their high content of potentially tooth-eroding silica.

Thomas Henry Huxley had worked on the same problem, and had proposed a similar evolutionary sequence, but Kovalevsky provided so much more documentation that primary credit for this seminal story fell to the Russian scientist (with Huxley's blessing). Kovalevsky and Huxley reconstructed the evolutionary sequence of horses as passing through a linear series of four successive stages, all based on European fossils (see the accompanying evolutionary tree, reproduced from Kovalevsky's German monograph of 1876): from the Eocene *Paleotherium,* to the early-Miocene three-toed *Anchitherium,* to the late-Miocene *Hipparion,* to *Equus,* the modern horse.

Kovalevsky had no misgivings about the factuality of his sequence. He wrote about the first step in his English monograph of 1874: "There can be, in my opinion, no reasonable doubt that the Horse descended from the *Paleotherium.*" He expressed equal assurance about step two in his French treatise of 1873: "In its skeleton, *Anchitherium* is a genus so intermediary, so transitional, that if the theory of transmutation were not already so firmly

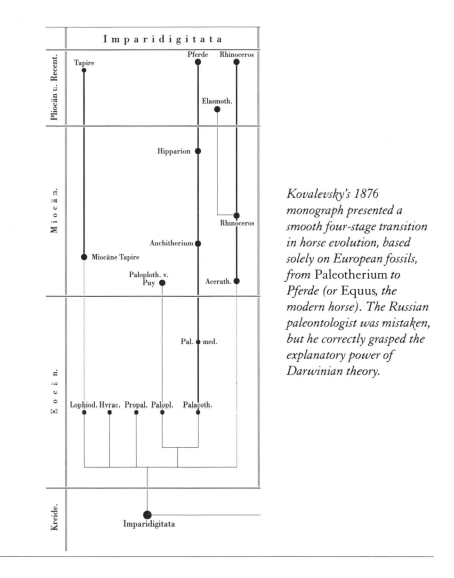

Kovalevsky's 1876 monograph presented a smooth four-stage transition in horse evolution, based solely on European fossils, from Paleotherium *to* Pferde *(or* Equus, *the modern horse). The Russian paleontologist was mistaken, but he correctly grasped the explanatory power of Darwinian theory.*

established, this genus would have provided one of the most important pillars of support" (my translations throughout). His German monograph of 1876 exuded confidence in the full sequence of four steps: "We have a form in the upper Eocene, which must surely be counted as an ancestor of horses; and we now have apparently persuasive data that this form, *Paleotherium medium,* evolved through the Miocene *Anchitherium* and *Hipparion* to modern horses." The French monograph of 1873 made the same claim in a more succinct and forceful manner: "There can be no reasonable doubt that the four forms—*Paleotherium medium, Anchitherium, Hipparion* and the horse—form a relationship of direct descent."

I would add a third point, more technical and professional and therefore not part of Kovalevsky's public reputation, to explain his high status among working paleontologists. Kovalevsky won kudos for his meticulous attention to descriptive detail. I have quoted the few general comments from his publications; the bulk of his text provides exhaustive information about every bump on every bone and every nubbin on every tooth. Moreover, Kovalevsky did not engage in such laborious work for the mindless and obsessive reasons that motivate many practitioners, but explicitly because he viewed this level of detail as a prerequisite for adequate documentation of such grand ideas as evolution, natural selection, and adaptation. Directly following his paean to Darwin from his English monograph (quoted earlier), and as an excuse for such flowery generality, he wrote: "The foregoing observations are intended only as a sort of apology for the somewhat minute osteological details into which it seemed to me necessary to enter in my description."

Kovalevsky began his French treatise of 1873 with his most explicit defense of detailed empiricism as a scientific method. He states that he was able to succeed in documenting the intermediary status of *Anchitherium* only because he had so many more fossils to study:

> I was able to obtain more complete material than any of my predecessors possessed. A monograph on *Anchitherium* now offers, following all the beautiful conquests of Darwinian theory, an irresistible charm for all evolutionary naturalists.

But Kovalevsky, true to the foremost ideal of objectivity, then denies that any a priori preference for evolution could have influenced his interpretations. In the best tradition of Sergeant Friday—"just the facts, ma'am"—he states that he had reached an inescapable conclusion by freeing his mind of preconceptions and letting the fossils speak in their undeniable factuality:

> Nevertheless, one must not think that I began this work with a preconceived aim. Quite to the contrary; I interrogated the material in an impartial manner, and I give the response that the fossils furnished to me.

With this brave claim, we come to a wonderful irony that raises the story of Vladimir Kovalevsky from an antiquarian tale for the heartstrings, to a lesson of central relevance for the practice of modern science. Kovalevsky developed a case study that remains preeminent as a triumphant tale of evolution documented in the fossil record. (Does any paleontological museum *not* include an evolutionary "line" of horses?) He expressed complete confidence in his sequence of four European genera. He won his reputation for unsurpassed care in meticulous and detailed description. He strongly espoused the classic doctrine of utter objectivity in observation, and he claimed that the fossils themselves had dictated an unimpeachable factual conclusion.

And Kovalevsky was wrong—for a clear and interesting reason that brings no discredit upon his work. He and Huxley had regarded the sequence of four European fossils as an unbroken lineage of direct transformation—that is, as a series of ancestors and descendants. They did not realize—and could not have done so in the absence of published evidence—that horses had evolved in America and had migrated several times into Europe. The three "ancestors" of Kovalevsky's sequence—*Paleotherium, Anchitherium,* and the European *Hipparion*—are all side branches that migrated to Europe and became extinct without issue in their new and peripheral home. (As a historical irony that would cost the Aztecs dearly and greatly facilitate the bloody schemes of the Conquistadores, horses then died out in America, leaving the Old World descendants of yet another migration as the source of all modern horses.) Kovalevsky's four genera did represent a temporal series of stages in a trend, but not—as he had asserted so confidently—a direct sequence of filiation. He had, so to speak, misidentified an indirect series of my paternal grandpa's brother, my mother's brother, and me, as a direct genealogy of my paternal grandpa, my father, and me.

The denouement occurred in an interesting manner when T. H. Huxley made his only trip to the United States to present an inaugural lecture for the opening of Johns Hopkins University, and to participate in several other events—including a lecture in New York on the evolution of horses—surrounding America's centennial celebration of 1876. He visited the American paleontologist O. C. Marsh at Yale and, in great excitement mixed with a bit of chagrin, saw enough beautifully transitional fossils to know that Europe had been a periphery, and America a breeding ground. Marsh later wrote of his magnificent show-and-tell:

> He [Huxley] then informed me that this was new to him, and that my facts demonstrated the evolution of the horse beyond question, and for the first time indicated the direct line of descent of an existing animal. With the generosity of true greatness, he gave up his own opinions in the face of new truth and took my conclusions.

Huxley scrapped his New York lecture, and hastily prepared a new version.

If science, as a stereotype proclaims, truly operated as an automatic process of objective documentation, then we should castigate Kovalevsky—for he firmly advertised his conclusions as factually driven and truly proven thereby. Moreover, if theories rest upon crucial cases, as another stereotype holds, then don't we call evolution itself into question by scratching Kovalevsky's horses from the race of life? Doesn't Kovalevsky then become an incubus rather than a hero—an unwitting foil for creationists rather than a preeminent student of evolutionary paleontology—because his error might give comfort to the enemy?

We need, instead, to reject these simplistic stereotypes and heed Marsh's wise statement about "the generosity of true greatness." The only universal attribute of scientific statements resides in their potential fallibility. If a claim cannot be disproven, it does not belong to the enterprise of science. New fields (like evolutionary paleontology in Kovalevsky's day) based on imperfect data (the fossil record) are especially subject to error—and brave scientists must do their best and risk the consequences, secure in the knowledge that corrected mistakes (however personally embarrassing to the perpetrator) bring as much clarity as valid discoveries.

Moreover, truly grand and powerful theories—evolution preeminently among them—do not and cannot rest upon single observations. Evolution is an inference from thousands of independent sources, the only conceptual structure that can make unified sense of all this disparate information. The failure of a particular claim usually records a local error, not the bankruptcy of a central theory. Kovalevsky mistook a series of collateral relatives for a true genealogical sequence, but the notion of genealogy itself does not fall thereby. If I mistakenly identify your father's brother as your own dad, you don't become genealogically rootless and created *de novo*. You still have a father; we just haven't located him properly.

With this tale of Kovalevsky's fruitful error, we come face to face with the interesting and conjoined issues of the relationship between fact and theory in science, and the phenomenon of being right for the wrong reason. Theory and fact are equally strong and utterly interdependent; one has no meaning without the other. We need theory to organize and interpret facts, even to know what we can or might observe. And we need facts to validate theories and give them substance. Kovalevsky made an eminently useful and rather wonderful mistake. He grasped the explanatory power of a new and correct theory that would reformulate all of biology—and he yearned to apply this idea to difficult but crucial data. He reached a premature conclusion that was only half wrong in misidentifying collateral relatives as direct ancestors. But he provided the first powerful example of a workable methodology (inferences about adaptation from anatomical changes in paleontological sequences) that can, with better data, document and support the most important theory subject to adjudication by the fossil record.

Lest anyone doubt the power of being right for the wrong reason, consider the central incident in the intellectual life of Kovalevsky's hero, Charles Darwin. Thanks to his obsessive habits of record keeping, we can reconstruct the sources of Darwin's eureka when, in early 1837, he recognized that evolution must be true. Darwin first learned that he had been wrong in allocating many small birds from the Galápagos Islands to several distantly related families. All, in fact, were finches. How could adjacent islands house separate species of such closely related birds? He then learned that the distinctive rhea (a large flightless bird) he had collected in southern Patagonia belonged to a new species (named *Rhea darwinii* in his honor by his ornithological consultant, John Gould). Why, Darwin wondered, did two so closely related birds share contiguous geographic territories, the ordinary rhea to the north, and Darwin's new species to the south?

Darwin pondered these two examples of geographic replacement among closely related modern birds. He then made a brilliant analogy: If both finches and rheas replace each other in space, then shouldn't temporal succession also occur in continuity—that is, by evolution rather than successive creation? Darwin had collected large fossil mammals in South America. He regarded one genus, later named *Macrauchenia,* as a relative of the modern guanaco. If the two rheas are maximally related and geographically contiguous, then the two camel-like creatures, the extinct *Macrauchenia* and the modern guanaco, as temporally contiguous, must also be joined by blood. In other words, temporal sequence must record evolutionary transformation. Eureka! Darwin jotted the key insight into a private notebook: "The same kind of relation that common ostrich [rhea] bears to Petisse [the new species, *Rhea darwinii*], extinct guanaco to recent; in former case position, in latter time."

What a portentous moment; what a crux in the history of human thought! But Darwin based his brilliant analogy, and his astoundingly correct general conclusion, on a flub in particulars—an error that he could not have recognized, and that did not affect the validity of his larger insight. South America had been an island continent for tens of millions of years before the Isthmus of Panama rose just a few million years ago (see chapter 20). Several independent orders of mammals had evolved on South America, though most perished following climatic changes and the influx of North American species after the isthmus formed. (The outstanding set of survivors, South America's grand indigenous group, includes sloths, armadillos, and anteaters of the unique order Edentata.)

One of these extinct and independent orders, the Litopterna, included creatures uncannily convergent upon unrelated mammals of other continents. (In the evolutionary phenomenon of convergence, distantly related groups evolve similar forms, as a result of independent adaptation to common environments. Ichthyosaurs are reptiles, and dolphins are mammals—but they both look and work like fishes.) One group of litopterns evolved an amazing resemblance to horses, including a parallel loss of toes, culminating in *Thoatherium,* the one-toed litoptern! Another group, represented by *Macrauchenia,* converged upon the New World camels. Since true camels later crossed the Isthmus of Panama and colonized South America (to survive today as llamas, guanacos, and alpacas), we can hardly blame Darwin for inferring a genealogical tie between modern guanacos and the remarkably similar but unrelated fossil *Macrauchenia.* Richard Owen, England's greatest anatomist and Darwin's friend in these early years (see chapter 6), had affirmed the link of *Macrauchenia* to modern guanacos.

Thus, at this most sublime moment in the history of biology, Darwin grasped the truth of evolution by basing a brilliant analogy on a flat-out factual error. Theories rarely arise as patient inferences forced by accumulated facts. Theories are mental constructs potentiated by complex external prods (including, in idealized cases, a commanding push from empirical reality). But the prods often include dreams, quirks, and errors—just as we may obtain crucial bursts of energy from foodstuffs or pharmaceuticals of no objective or enduring value. Great truth can emerge from small error. Evolution is thrilling, liberating, and correct. And *Macrauchenia* is a litoptern. The fossil record provides our best direct evidence for evolution at large scales. And European fossil horses are collateral relatives, not ancestors of modern *Equus*.

As Mrs. Julia Ward Howe—the enduringly famous member of the couple—wrote, we may obtain our inspiration from a full range of sources, both "in the beauty of the lilies," and "where the grapes of wrath are stored" (to probe the entire botanical spectrum). Any inspiring light can indicate a path in the darkness of nature's complexity. "I have read His righteous sentence by the dim and flaring lamps. His truth is marching on."

III
HUMAN
PREHISTORY

8

UP AGAINST A WALL

WE ARE, ABOVE ALL, A CONTENTIOUS LOT, UNABLE TO AGREE ON MUCH OF anything. Alexander Pope caught the essence of our discord in a couplet (though modern technology has vitiated the force of his simile):

> *'Tis with our judgments as our watches, none*
> *Go just alike, yet each believes his own.*

Most proclamations of unanimity therefore convey a fishy odor—arising either from imposed restraint ("elections" in dictatorial one-party states), or comedic invention to underscore an opposite reality (as when Ko-Ko, in Gilbert and Sullivan's *Mikado,* reads a document signed by the Attorney-General, the Lord Chief Justice, the Master of the Rolls, the Judge Ordinary, and the Lord Chancellor—and then proclaims, "Never knew such unanimity on a point of law in my life." But the document has been endorsed by only one signatory—for Pooh-Bah holds all the aforementioned titles!).

Paleontologists probably match the average among human groups for levels of contentiousness among individuals (while students of human pre-history surely rank near the top, for this field contains more practitioners than objects for study, thus breeding a high level of acquisitiveness and ter-ritoriality). Yet one subject—and only one—elicits absolute unanimity of judgment among students of ancient life, though for reasons more visceral than intellectual. Every last mother's son and daughter among us stands in reverent awe and amazement before the great cave paintings done by our ancestors in southern and central Europe between roughly thirty thousand and ten thousand years ago.

If this wonderment stands as our only point of consensus (not confined, by the way, to scientific professionals, but shared with any member of *Homo sapiens* possessing the merest modicum of curiosity about our past), please don't regard me as a Scrooge or a Grinch if I point out that our usual ratio-nale for such awe arises from a pairing of reasons—one entirely appropri-ate, and the other completely invalid. For I don't impart this news to suggest any diminution of wonder, but rather to clear away some conceptual bag-gage that, once discarded, might free us to appreciate even more fully this amazing beginning of our most worthy institution.

For the good reason, we look at the best and most powerful examples of this art, and we just know that we have fixed a Michelangelo in our gaze. Comparisons of this sort seem so obvious, and so just, that they have become a virtual cliché for anyone's description of a first reaction to a wonderfully painted cave wall. For example, in describing his emotional reaction to the newly discovered cave of Chauvet—the source of eventual dénouement for this essay as well—a noted expert wrote: "Looking closely at the splendid heads of the four horses, I was suddenly overcome with emotion. I felt a deep and clear certainty that here was the work of one of the great masters, a Leonardo da Vinci of the Solutrean revealed to us for the first time. It was both humbling and exhilarating."

For the bad argument, our amazement also arises for a conceptual rea-son added to our simple (and entirely appropriate) visceral awe. We are, in short, surprised, even stunned, to discover that something so old could be so sophisticated. Old should mean rudimentary—either primitive by greater evolutionary regress toward an apish past, or infantile by closer approach to the first steps on our path toward modernity. (These metaphors of grunting coarseness or babbling juvenility probably hold about equal sway in the formation of our prejudices.) As we travel in time down our own evolutionary tree, we should encounter ever-older ancestors of ever-decreasing mental capacity. The first known expressions of representational art should therefore be crude and primitive. Instead, we see the work of a primal Picasso— and we are dumbstruck.

I dedicate this essay to tracing the prevalence of this view in the lifework of the two greatest scholars of Paleolithic cave art. I shall then argue that this equation of older with more rudimentary both violates the expectations of evolutionary theory when properly construed, and has now also been empirically disproven by discoveries at Chauvet and elsewhere. I shall then suggest that the more appropriate expectation of maximal sophistication for this earliest art should only increase our appreciation—for we trade a false (if heroic) view of ever-expanding triumph for a deeply satisfying feeling of oneness with people who were, biologically, fully us in circumstances of maximal distance, both temporal and cultural, from our current lives.

(No species of punditry deserves more ridicule than the art form known as Monday-morning quarterbacking or backseat driving—the "I told you so" of the nonparticipant. This essay veers dangerously toward such an unworthy activity. I am, after all, a paleontologist and expert on land snails, not an art historian or a student of human culture. What right do I have to criticize the monumental and lifelong efforts of the Abbé Henri Breuil and André Leroi-Gourhan, the most learned and prolific of true devotees? In defense, I would say, first, that honorable errors do not count as failures in science, but as seeds for progress in the quintessential activity of correction. No great and new study has ever developed without substantial error, and we need only cite a famous line from Darwin: "False facts are highly injurious to the progress of science, for they often endure long; but false views, if supported by some evidence, do little harm, for every one takes a salutary

pleasure in proving their falseness." In reverse of Marc Antony, I have come to praise Breuil and Leroi-Gourhan, not to bury them. Second, the perspective of cognate fields can often bring light to neighboring disciplines too set in favored ways. I therefore speak from my close vicinity of evolutionary theory to point out—as many have before me, and on the same basis—that conventional expectations have no sanction in our current understanding of evolution, and therefore represent lingering prejudices that we might wish to reassess and then choose to discard.)

The general title of "ice-age" or "Paleolithic" (literally, Old Stone Age) art has been applied to the great variety and geographic spread of works in two major categories—smaller and movable objects usually called "portable" (the so-called Venus figurines, the deer, horses, and other animals carved in bone or ivory on disks, plaques, and spear-throwers, for example); and the engravings and paintings on cave walls (and now from a few open sites as well), dubbed "parietal." (A *paries* is a Latin wall; if you belong to my generation, you will remember, now with amusement but then with utmost frustration, the parietal hours of college dorms, when members of the opposite sex had to return within the walls of their own rooms, and not remain in yours.) European portable art extends from Spain to Siberia; parietal art has been found mostly in France and northern Spain, with a few Italian sites, and perhaps others even more distant. (Decorated caves in other genres, but perhaps of equal or even greater age, have been found in many other areas of the world, from Africa to Australia.)

Current radiocarbon dates (from charcoal) of paintings in parietal art span a range from 32,410 years B.P. (before the present) at Chauvet, to 11,600 years B.P. at Le Portel. This period corresponds to the occupation of Europe by our own species, *Homo sapiens* (often called "Cro-Magnon" in this incarnation to honor a French site of first discovery. Remember that the Cro-Magnon people are us—by both bodily anatomy and parietal art—not some stooped and grunting distant ancestor). The immediately earlier inhabitants of Europe, the famous Neanderthal people, did not (so far as we know) produce any representational art. Neanderthals overlapped Cro-Magnon in Europe, probably into the time of Cro-Magnon's early parietal art. This striking cultural difference reinforces the opinion that Neanderthal and Cro-Magnon were two separate, albeit closely related, species, and not endpoints of a smooth evolutionary continuum. On this view, Neanderthal died out, while Cro-Magnon continues as us (see chapter 10).

Two subjects have long dominated the theoretical discussion on parietal art: function and chronology. The two greatest scholars in this field—the Abbé Henri Breuil and André Leroi-Gourhan—differed profoundly in their views about function, but (somewhat paradoxically) agreed substantially in their proposals about chronology.

Of the several French priests who have become distinguished students of ancient life (in a land where both Catholic traditions and general intellectual commitments favor harmony between the different domains of science and religion—see chapter 14), Pierre Teilhard de Chardin surely won the most fame, while the Abbé Henri Breuil may well have done the best work. Breuil, a talented artist, spent nearly sixty years copying figures from cave walls (at a time when photography yielded poor images in such subterranean conditions), and then comparing the results in his compendia of drawings. He traced directly from cave walls whenever possible, and with utmost care at all times (drawings are not inherently more subjective than photographs). But he sometimes had to work by difficult and indirect methods. He could not, for example, press paper against the famous painted ceiling of Altamira because any direct contact detached the pasty pigment used by Paleolithic artists. Therefore, positioned like Michelangelo under the Sistine Chapel ceiling, he lay on his back, cushioned on soft sacks of ferns, while holding his paper as close to the roof as possible, and making imperfect sketches.

As he drew the animals one by one, Breuil tended to read their meaning in the same piecemeal fashion—that is, as individuals rather than parts of integrated compositions. He held that the paintings functioned as a kind of "hunting magic" to make game plentiful (if you draw it, it will come), or to ensure success in the kill (game animals are often painted with wounds and spear holes). Breuil wrote in his summary book of 1952:

> Here, for the first time, men dreamed of great art and, by the mystical contemplation of their works, gave to their contemporaries the assurance of success in their hunting expeditions, of triumphs in the struggle against the enormous pachyderms and grazing animals.

In the next generation, André Leroi-Gourhan, director of the Musée de l'Homme in Paris, approached the same subject of meaning from the maximally different perspective of his "card-carrying" membership in one of the major intellectual movements of our century—French structuralism as embodied in the work of anthropologist Claude Lévi-Strauss. This form of structuralism searches for timeless and integrative themes based on dichotomous divisions that may record much of nature's reality, but mostly reflect the brain's basic mode of operation. Thus, we separate nature from culture (the raw and the cooked, in Levi-Strauss's terminology), light from darkness, and, above all, male from female.

Leroi-Gourhan therefore viewed each cave as an integrated composition, a sanctuary in which the numbers and positions of animals bore unified meaning within a scheme set by the primary duality of male and female. Each animal became a symbol, with a primary division between horses as male and bisons as female. He also interpreted abstract signs and artifacts as sexually labeled, with spears (for example) as male, and wounds as female. He viewed the cave itself as fundamentally female, thus requiring a definite positioning and grouping of male symbols. Led by this theory, Leroi-Gourhan treated each cave as a unity and compiled extensive statistical tables of numbers and locations—in maximal contrast with Breuil's concentration on each animal in and for itself. Leroi-Gourhan wrote:

> Clearly, the core of the system rests upon the alternation, complementarity, or antagonism between male and female values, and one might think of "a fertility cult." . . . There are few religions, primitive or evolved, that do not somewhere involve a confrontation of the same values, whether divine couples such as Jupiter and Juno are concerned, or principles such as yang and yin. There is little doubt that Paleolithic men were familiar with the division of the animal and human world into two opposite halves, or that they supposed the union of these two halves to govern the economy of living beings . . . Paleolithic people represented in the caves the two great categories of living creatures, the corresponding male and female symbols, and the symbols of death on which the hunters fed. In the central area of the cave, the system is expressed by groups of male symbols placed around the main female figures, whereas in the other parts of the sanctuary we find exclusively male representations, the complements, it seems, to the underground cavity itself.

And yet, despite their maximal ideological difference on the function and meaning of cave art, Breuil and Leroi-Gourhan maintained agreement on the second great subject of chronology. To be sure (and as we shall see), these two great scholars diverged on many particulars, but they shared an unswerving and defining conviction—a kind of central and unshakable faith—that the chronology of cave art must record a progression from crude and simple beginnings to ever more refined and sophisticated expression. In so doing, these scholars could assimilate the earliest known history of representational art to the classic myths and sagas of Western culture—the hero's birth, his first faltering steps, his rise to maturity, his triumph and dominion, and, ultimately, his tragic fall (for both Breuil and Leroi-Gourhan included a final stage of degeneration after the ice retreated and the game dispersed).

In regarding a progressive chronology as a consummation so devoutly to be wished, Breuil and Leroi-Gourhan had complex motives. They were, no doubt and in large part, simply caught up in conventional modes of thinking, so deep and so automatic in our culture that such views rarely bubble to a conscious surface where they might be questioned. But an important technical reason also drove both scholars to such a hope. Layers of sediment can often be dated by various means, now conventional. But a cave is a hole in the ground; how can you specify the age of a cavity? (You might ascertain the age of rocks forming the cave wall, but these dates bear no relationship at all to the age of the cave as a hole.) How, then, can you know the time of a prehistoric painting or engraving on a cave wall? (Today we can date the pigments by carbon-14 and other methods, particularly the charcoal used to draw black lines, but Breuil had no access to such techniques at all, and the carbon-14 methods of Leroi-Gourhan's time required so much material to obtain a date that entire paintings would have been sacrificed—a procedure that no one, quite properly, would ever sanction.)

The only hope for dating therefore inhered in the paintings themselves—in the search for an internal criterion that could order this earliest art into a chronological sequence. Breuil struggled mightily to establish such

an order by superposition—that is, by studying paintings drawn over earlier paintings. He succeeded to some extent, but technical problems proved
too daunting for a general solution. You cannot always specify the sequence
of overlap on an essentially flat surface; moreover, even if you can, the painting on top could have been executed the next day, or a thousand years after,
the one below. Leroi-Gourhan contrasted the ease of dating portable objects
found in strata with difficulties for paintings on cave walls: "A reindeer
incised on a small plaque, found in a layer that also yielded hundreds of
flints, is often easy to date, but a mammoth painted on a cave wall three feet
or more above the ground is cut off from all chronological clues."

Both scholars therefore turned to the venerable technique of art historians of later times—the analysis of styles. But a problem of circular reasoning now intrudes, for we need a source of evidence separate from the
paintings themselves. We can place Michelangelo's style in the sixteenth century, and Picasso's in our own, because we have independent evidence about
dates from a known historical record. But nothing either in abstract logic or
pictorial necessity dictates that one form of mannerism must be four hundred years old, while another style of cubism could only emerge much later.
If we had absolutely no other evidence but Michelangelo's *Last Judgment*
and Picasso's *Guernica*—no texts, no contexts, no witnesses—we could not
know their temporal order.

In such a context of abysmally limited information—the situation faced
by Breuil and Leroi-Gourhan—we must try to construct a theory of stylistic
change that might establish a chronological sequence by internal evidence.
(If we could say, for example, that realism must precede abstraction, then we
could place Michelangelo before Picasso by internal criteria alone.) I don't
fault these scholars for seeking such a theory of stylistic change—for how
else could they have proceeded, given the limitations? But I am intrigued
that they fell back so easily and so uncritically—almost automatically, it
might seem—upon the most conventional form of progressivist mythology:
a chronology ordered by simple to complex, or rude to sophisticated.

I can better grasp Breuil's attraction to the legend of progress. He was, after all, a child of the late nineteenth century—the great age of maximal faith in human advance, especially in Western nations at the height of their imperial and industrial expansion (the ravages of World War I ended this illusion for many, though not, apparently, for Breuil). But Leroi-Gourhan's assent is more puzzling, for his philosophical commitment to structuralism led him to view the symbolic ensemble of each cave as the expression of an unvarying human psyche, with its dualistic contrasts of male and female, danger and safety, and so forth.

In fact, and in several interesting passages, Leroi-Gourhan addressed this issue directly. He acknowledges that structuralism does lead to a hypothesis of unvarying form and function for caves as sanctuaries through-out the history of Paleolithic parietal art. But given this constancy of structure, he then argues, how could we untangle chronology except by the hope (and expectation) that styles used to paint the constant symbols will change in a systematic way through time? A bison may always represent the female moiety, and may always occupy the same position within a cave—but artists may learn to paint bisons better through time. Leroi-Gourhan writes:

> The same content persists from first to last. The pairing of animal species with signs appears in the Aurignacian [the first period of cave art] and disappears in the terminal Magdalenian [the last period]. Consequently, the ideological unity of cave art rules out the guideposts that it might provide for us had there been changes in the basic themes. Only variations in the representation of this uniform subject matter are discernible in the course of a stylistic study.

Parietal art includes a complex array of figures and signs. The figures mostly depict the large mammals of ice-age Europe (various deer, horses, bison, mammoths, rhinos, lions, and several others), but we also find occasional humans (and the more frequent, and wonderful, handprints, often stenciled by placing a hand against the wall and blowing paint around it with some kind of Paleolithic spray can). In another category, rarely given as much attention but surely surpassing the animal figures in number (and perhaps in interpretive importance), a large variety of signs and symbols festoon the walls—some identifiable as pictures of weapons or body parts (often sexual), others as geometric forms, and still others quite mysterious.

In the progressivist chronologies of Breuil and Leroi-Gourhan, figures and signs show a superficially opposite directionality. Figures begin with crude and simple outlines and progress to more supple and complex realism, complete with dimensionality and perspective. Signs, on the other hand, become simpler and more symbolic, with identifiable pictures (of vulvas, for example) evolving to less variable, more symbolic, and often highly simplified geometric representation. Leroi-Gourhan wrote: "The animal figures . . . show a development in form towards a more and more precise analysis. The geometricization of the signs in contrast with the character of the animal figures is one of the interesting aspects of research into the meaning of the designs."

But these apparently opposite directions of change for figures and symbols really represent—as Breuil and Leroi-Gourhan repeatedly emphasized—different facets of the same overall theme of progress as the basis of chronology. In painting figures, the artists were trying to do better in representing the animals themselves   and the supposed sequence of styles marks their continual improvement. But, in drawing signs, the same artists were knowingly developing a system of symbols—and symbols gain universality and meaning by becoming more abstract and reduced to a geometric essence. After all—and the analogy was not lost on these scholars—most alphabets derived their letters as simplified pictures of objects (while the same argument applies with even more force to the evolution of such character systems as Chinese).

Breuil initially proposed a system of five stages in a single sequence of greater realism and complexity for figures (his papers of the early 1900s make fascinating reading). He later developed his famous theory of two successive cycles, each with a complete history of progress to a pinnacle, followed by late decline. (Breuil continued to hold, despite mounting evidence to the contrary, that the first cycle could be recognized by drawings of animals in "twisted perspective"—that is, with more than one plane represented, as in a bison with a body seen from the side, but with a face pointing forward.)

Breuil's two schemes are not so different as they appear—for his first notion of five sequential steps included a period of decline in the middle. I was particularly struck by his adjectives of judgment in supposedly objective descriptions of the stages. In an early article of 1906, he marks the animals of Stage 3 (later to become the decline at the end of the first cycle) as "of a deplorable design, and with a disconcerting lack of proportion." He then praises the recovery of Stage 4 as one might describe a Renaissance artist trying to re-create the lost glories of an ancient Greece: "The artists sought to rediscover the model lost in the preceding stage. They obtained this result by polychromy [figures of more than one color]. These paintings are timid at first . . ." Breuil concluded his paper by stating: "Paleolithic art, after an almost infantile beginning, rapidly developed a lively way of depicting animal forms, but didn't perfect its painting techniques until an advanced stage."

Leroi-Gourhan, in contrast, developed a theory of four successive stages in a single series. But his sequence of progress scarcely differed from Breuil's—though the older scholar wanted to run the story twice. Both schemes began with immobile animals stiffly carved in crude outlines with no interior coloration, and moved on to ever more accurate images, drawn with a much better feel for mobility, rendered in better perspective, and more richly colored. (The later artists, Leroi-Gourhan believed, reached such a state of perfection that their art stagnated a bit at the end, becoming rather academic in replication of excellence.)

Mario Ruspoli, a disciple of Leroi-Gourhan, epitomized the theory well in his 1986 book *The Cave of Lascaux*. "From the earliest images onward, one has the impression of being in the presence of a system refined by time . . . The development of Paleolithic cave art may be summed up as 15,000 years of apprenticeship followed by 8,000 years of academicism."

Leroi-Gourhan recognized the essential similarity of his view with the

earlier theories of Breuil. After a detailed (though respectful) critique of
Breuil, and an extensive compendium of their particular differences, Leroi-
Gourhan acknowledges the fundamental similarity in their common con-
cept of progress as the key to a chronology of Paleolithic art:

> The theory . . . is logical and rational: art apparently began with
> simple outlines, then developed more elaborate forms to achieve
> modeling, and then developed polychrome or bichrome painting
> before it eventually fell into decadence.

This progressivist theory of increasingly complex and supple realism in
Paleolithic painting dominated the field for decades. Writing of Leroi-
Gourhan's four-stage theory, Brigitte and Gilles Delluc (in Ruspoli's book,
cited previously) state simply: "The classification was fairly soon adopted by
everyone." And yet, I think everyone now realizes that the hypothesis of
progressivism in Paleolithic art cannot hold. The march to greater and more
complex realism doesn't make any sense theoretically, and has now been disproven empirically at Chauvet and elsewhere.

Theoretical dubiety. I don't want to use this essay as one more rehearsal
for my favorite theme that Darwinian evolution cannot be read as a theory
of progress, but only as a mechanism for building better adaptation to
changing local environments—and that the equation of evolution with
progress represents our strongest cultural impediment to a proper understanding of this greatest biological revolution in the history of human
thought. Still, I can't help pointing out that this prejudice must underlie the
ready proposition and acceptance of such a manifestly improbable notion as
linear progress for the history of parietal art from thirty thousand to ten
thousand years ago.

But why do I label the progressivist hypothesis of Breuil and Leroi-
Gourhan as "manifestly improbable"? After all, humans did evolve from
apish ancestors with smaller brains and presumably more limited mental
capacities, artistic and otherwise. So why shouldn't we see progress through
time?

The answer to this query requires a consideration of proper scale. The twenty-thousand-year span of known parietal art does not reach deep into our apish ancestry (where a notion of general mental advance could be defended). The earliest parietal art lies well within the range of our current species, *Homo sapiens*. (By best estimates, *Homo sapiens* evolved in Africa some 200,000 years ago, and had probably migrated into the Levant [if not into Europe proper] by about ninety thousand years ago.) Therefore, the painters of the first known parietal art were far closer in time to folks living today than to the original *Homo sapiens*.

But a progressivist critic might still retort: "Okay, I now understand that we are only discussing a sliver of human history, not most of the whole story since our split from the common ancestor of chimpanzees and us. But the trend of the whole should also be manifest within the shorter history of individual species, for evolution should move slowly and steadily to higher levels of mentality." Herein lies the key prejudice underlying our uncritical acceptance of the progressivist paradigm for the history of art. It just feels "right" to us that the very earliest art should be primitive. Older in time should mean more and more rudimentary in mental accomplishment.

And here, I think, we make a simple (but deep and widespread) error. Apparently similar phenomena of different scale do not become automatically comparable, but often (I would say usually) differ profoundly. Changes *between* species in an evolutionary sequence represent a completely different phenomenon from variation (spatial or temporal) *within* a single species. Humans have bigger brains than ancestral monkeys; these monkeys have bigger brains than distantly ancestral fishes. This increase in brain size does record a great gain in mental complexity. But a correlation of size and smarts across species does not imply that variation in brain size among modern humans also correlates with intelligence. In fact, normal adults differ in brain size by as much as 1,000 cubic centimeters, and no correlation has ever been found between size and intelligence (the average human brain occupies about 1,300 cc of volume).

Similarly, while evolution obviously produces change between one species and the next in a sequence of descent, most individual species don't alter much during their geological lifetimes. Large, widespread, and successful species tend to be especially stable. Humans fall into this category, and the historical record supports such a prediction. Human bodily form has not altered appreciably in 100,000 years. As I stated earlier, the Cro-Magnon cave painters are us—so why should their mental capacity differ from ours? We don't regard Plato or King Tut as dumb, even though they lived a long time ago. Remember that the distance from Plato to the parietal painters spans far less time than the interval separating these painters from the first *Homo sapiens.*

But defenders of progressivism in parietal art might still fall back upon one potentially promising argument: cultural change differs profoundly from biological evolution. We can admit biological stability and still expect an accumulative and progressive history of art or invention. The road has been both long and upward from Jericho and some scratch farming to New York City and the World Wide Web.

Fair enough in principle—but, again, the known timing precludes such an argument in practice. I will admit that if we happen to catch art at the very beginning, we would not expect full sophistication right away. But the oldest known parietal art, at thirty thousand years ago, lies well into the history of *Homo sapiens* in Europe—far closer to us today than to the first invasion from Africa. I don't know why earlier art hasn't been found (perhaps we just haven't made the discovery yet; perhaps people only moved into areas with caves at a much later time). I doubt that Ugh, the first Cro-Magnon orator, spoke in truly dulcet tones. But we surely don't regard Pericles as worse than Martin Luther King, Jr., just because he lived a few thousand years ago. Phidias doesn't pale before Picasso, and no modern composer beats Bach by mere virtue of residence in the twentieth century. Please remember that the first known Cro-Magnon artist, at thirty thousand years ago, stands closer to Pericles and Phidias than to Ugh, the orator, and Ur, the very first painter. So why should parietal art be any more primitive than the great statue of Athena that once graced the Parthenon?

As a final point, why should areas as distant as southern Spain, northeastern France, and southeastern Italy go through a series of progressive stages in lockstep over twenty thousand years? Regional and individual variation can swamp general trends, even today in a world of airplanes and televisions. Why did we ever think that evolution should imply a primary signal of uniform advance?

This general line of criticism has been well articulated by Paul G. Bahn and Jean Vertut in their 1988 book *Images of the Ice Age*. (I am pleased that they found our paleontological theory of punctuated equilibrium useful in constructing their critique.)

> The development of Paleolithic art was probably akin to evolution itself: not a straight line or ladder, but a much more circuitous path—a complex growth like a bush, with parallel shoots and a mass of offshoots; not a slow, gradual change, but a "punctuated equilibrium," with occasional flashes of brilliance . . . Each period of the Upper Paleolithic almost certainly saw the coexistence and fluctuating importance of a number of styles and techniques, . . . as well as a wide range of talent and ability . . . Consequently, not every apparently "primitive" or "archaic" figure is necessarily old (Leroi-Gourhan fully admitted this point), and some of the earliest art will probably look quite sophisticated.

Empirical disproof. Theoretical arguments may be dazzling, but give me a good old fact any time. The linear schemes of Breuil and Leroi-Gourhan had been weakening for many years as new information accumulated and old certainties evaporated. But one technical advance truly opened the floodgates. Thanks to a new method of radiocarbon dating—called AMS, for Accelerator Mass Spectrometry—only tiny amounts of charcoal need now be used, and paintings may therefore be analyzed without removing significant material.

In late 1994, three French explorers discovered a wonderful new site, now called Chauvet Cave. The animals at Chauvet, particularly the magnificent horses and lions, match anything else in Paleolithic art for sophistication and accuracy. But the radiocarbon dates, multiply repeated and presumably accurate, give ages in excess of thirty thousand years—making Chauvet the oldest of all known caves with parietal art. If the very oldest includes the very best, then our previous theories of linear advance must yield. In his epilogue to a gorgeous book, published in 1996, on this new site, Jean Clottes, a leading expert on Paleolithic art, writes:

> The subdivision of Paleolithic art proposed by Leroi-Gourhan, in successive styles, must be revised. His Style I, in which Chauvet Cave should be placed, was defined as archaic and very crude without any definite mural depictions, and is obviously no longer adequate. We now know that sophisticated techniques for wall art were invented . . . at an early date. The rendering of perspective through various methods, the generalized use of shading, the outlining of animals, the reproduction of movement and reliefs, all date back more than 30,000 years . . . This means that the Aurignacians, who coexisted with the last Neanderthals before replacing them, had artistic capabilities identical to those of their successors. Art did not have a linear evolution from clumsy and crude beginnings, as had been believed since the work of the Abbé Henri Breuil.

Let us not lament any lost pleasure in abandoning the notion that we now reside on an ever-rising pinnacle of continuous mental advance, looking back upon benighted beginnings. Consider instead the great satisfaction in grasping our true fellowship with the first known Paleolithic artists. There but for the grace of thirty thousand additional years go I. These paintings speak so powerfully to us today because we know the people who did them; they are us.

In a famous paradox, Francis Bacon wrote: *antiquitas saeculi, juventus mundi* (or, roughly, "the old days were the world's youth"). In other words, don't think of the Paleolithic as a time of ancient primitivity, but as a period of vigorous youth for our species (while we today must represent the graybeards). Paleolithic art records our own early age, and we feel a visceral union with the paintings of Chauvet because, as Wordsworth wrote, "the child is father of the man." But we should also note the less frequently cited first verse of his poem:

> *My heart leaps up when I behold*
> *A rainbow in the sky;*
> *So was it when my life began;*
> *So is it now I am a man.*

We have loved the rainbow for thirty thousand unbroken years and more. We have struggled to depict the beauty and power of nature across all these ages. The art of Chauvet—and Lascaux, and Altamira, and a hundred other sites—makes our heart leap because we see our own beginnings on these walls, and know that we were, even then, worthy of greatness.

9

A LESSON FROM THE

OLD MASTERS

THE MOST FAMOUS LITERARY TALE OF A HUMP INVOKES AN EVOLUTIONARY theme of sorts. "In the beginning of years," Rudyard Kipling tells us in his *Just So Stories,* "when the world was so new and all, and the animals were just beginning to work for Man, there was a camel, and he lived in the middle of a Howling Desert because he did not want to work." Instead, when urged to service by the horse, dog, and ox, the recalcitrant camel merely snorted, "Humph." So the most powerful of resident Djinns, converting utterance to substance, put a hump on the camel's back to make up for three lost days of work at the beginning of time: "'That's made a-purpose,' said the Djinn, 'all because you missed those three days. You will be able to work now for three days without eating, because you can live on your humph.'"

Kipling ripped off the camel to preach a sermon for children about old-fashioned virtues of work, and the perils of idleness—for his accompanying poem abandons the charm of the tale itself for a heavy moral disquisition couched in doggerel:

The Camel's hump is an ugly lump
Which well you may see at the Zoo;
But uglier yet is the hump we get
from having too little to do.

I think that we owe nature a favor in return to expiate this exploitation of a long-standing evolutionary product, developed without the slightest human influence, and presumably long before our origin. So I too have a tale of a hump to tell—but for a different animal and an opposite purpose. In Kipling's version, the camel develops a hump in order to serve a human master diligently. In the story of this essay, we discover the existence of a hump only because ancient humans painted pictures of a feature that no conventional evidence of the fossil record could ever have revealed. I hope that nature will accept this trade: we rip off a well-known hump to construct a moral fable of dubious merit (Kipling's camel), but our ancestors restore another by providing the only possible evidence for a hump that would otherwise have disappeared into the maw of lost history (the Irish Elk of this essay).

We know that certain mammals, from camels to Quasimodo, have humps. Deer, however, do not grow humps—although large deer with big antlers (moose, in particular) often develop a broadly raised area on their backs, in the shoulder region where forelegs meet backbone. But the deer with the largest antlers of all time, the extinct (and misnamed) Irish Elk, did evolve a prominent hump—a wonderful fact that we can only know because human artists painted these giant deer on cave walls. A hump, as fatty tissue, does not fossilize.

Megaloceros giganteus, the so-called Irish Elk, surely heads the hit parade of extinct deer. In a famous quip, Voltaire remarked that the Holy Roman Empire had been misnamed in all attributes—for this amalgam of largely Germanic lands in central Europe was neither holy, Roman, nor an empire. Similarly, the Irish Elk was neither exclusively Irish nor an elk. This species lived in temperate climates throughout Europe and western

Asia (with close relatives in Siberia and China), from about 400,000 years ago to a last record in Ireland at 10,600 years B.P. (before the present). The Irish epithet derives from the superb preservation and frequent occurrence of Irish specimens, buried (and hermetically sealed) in sediments beneath layers of peat in the island's numerous bogs. A cottage industry developed in the nineteenth century for excavating and selling specimens to museums and collectors throughout the world—hence the identification with Ireland. (A 1994 article by Adrian M. Lister provides a summary of virtually all science and lore about Irish Elks. I have also studied this species extensively, and have published both technical accounts see my 1974 article cited in the bibliography—and general articles, including the very first essay that I ever wrote in this series, now spanning eight volumes and more than 250 essays.)

The "elk" misnomer has a more complex history. Early scientists thought that the Irish fossils might represent the same species as the American moose, then poorly known. Moose, in Europe, are called elk—hence the confusion. In any case, since *Megaloceros* is not a moose, the common name makes no sense. In this essay, I shall follow the practice of all current experts on these fossils, and refer to *Megaloceros* as the "giant deer."

Giant deer had large bodies, about equal in size to those of modern moose, although slightly exceeded by a fossil deer species or two. But the antlers of *Megaloceros*—the source of celebrity for the genus—hold all records for size and weight. Growing outward from the head, essentially at right angles to the body axis, these large palmated antlers (platelike rather than sticklike) could reach a span of up to thirteen feet from tip to tip, and a weight of one hundred pounds. When we recognize that male giant deer shed and regrew these structures annually (females grew no antlers), our wonder at the energetic drain can only increase.

In the light of this essay's focus on earliest human interactions with giant deer, I note that the history of scientific discussion about *Megaloceros* has always centered on questions of potential human contact with such a bizarre and fascinating creature. Two issues dominated the early literature.

1. IS THE GIANT DEER, OR ANY SPECIES, FOR THAT MATTER, TRULY EXTINCT? In the eighteenth century, as the Linnaean approach to classification became codified, and as the nascent science of geology began to reveal the earth's great age, a major debate arose among European naturalists: Could an entire species become extinct? Many leading naturalists rejected the possibility, either on traditional creationist grounds (for a hole would then be left in a system of relationships ordained as permanent and complete by an omniscient God), or by arguments derived from early forms of evolutionary thought (in Lamarck's system, for example, species maintained too much adaptive flexibility to die, though they could transform to higher states).

But if species couldn't die, where was the animal that, in ages past, left such magnificent antlers under the Irish peat bogs? Some scientists believed that the uncharted forests of Canada might still house the giant deer, perhaps in the degenerated form of the smaller-antlered American moose. (As mentioned previously, this conjecture led to the false name of "Irish Elk" for the giant deer.)

This debate ramified into a set of interesting byways. On the political front, a full-time statesman and sometime paleontologist named Thomas Jefferson blasted the great French naturalist Georges Buffon for his claim that all American species must be smaller and degenerated versions of European forms (including the American moose as a demoted giant deer). As his *touché,* Jefferson wrote a paper on the fossil claw of a giant lion, surely larger than any Old World counterpart. Unfortunately, the claw actually belonged to a large ground sloth—showing once again that neither patriotism nor morality should be staked upon the uncertain facts of nature.

On the artistic front, Britain's finest painter of animals, George Stubbs, did a portrait of the Duke of Richmond's yearling bull moose, the first to enter Britain. This work, executed in 1770, depicts the young moose on a mountain ledge, with storm clouds gathering in the background, and a pair of adult antlers lying in the foreground. The painting has long been celebrated, but the circumstances of composition have only recently come to light. The work was commissioned by the great Scottish medical anatomist

Stubbs's painting of a moose, done as a contribution to the debate about whether the Irish elk still lived in America.

William Hunter as part of a project (never published) to determine whether or not the American moose might represent the same species as the fossil Irish giant deer—hence Stubbs's depiction of *adult* antlers in the foreground! (See W. D. Ian Rolfe's 1983 article, cited in the bibliography.)

Proponents for extinction slowly gained the upper hand as further exploration, including the expedition of Lewis and Clark, encountered no living *Megaloceros*—while moose dropped from the running as their differences from giant deer become more apparent. Georges Cuvier, Europe's premier anatomist and founder of modern vertebrate paleontology, provided a final resolution in 1812, when he published his four volume

Recherches sur les ossemens fossiles (Researches on Fossil Bones) and proved both the fact of extinction in general, and the death of the giant deer in particular. Speaking with customary force, Cuvier wrote of the giant deer:

> Here is the most famous of all fossil ruminants, the one that naturalists regard, with greatest unanimity, as lost from the earth . . . It is certain that the Irish antlers could not belong either to the moose or the reindeer . . . This [fossil] species could not possibly be confused with any large [modern] deer on any continent.

2. DID HUMANS EVER INTERACT WITH GIANT DEER? Once the fact of extinction had been settled, scientists turned their attention to the timing and manner of dying. Proponents of human interaction with giant deer suffered a major setback in the key Irish localities, for humans did not reach the Emerald Isle (or at least did not leave any known bones or artifacts to indicate their presence) until well after the demise of *Megaloceros.* What, then, of the giant deer's large European range? Our Neanderthal cousins and, later, our Cro-Magnon ancestors certainly overlapped the giant deer in time, but did humans ever interact with *Megaloceros,* or did we share territory in mutual ignorance, like Longfellow's ships passing in the night?

Giant deer are never common in continental European fossil beds. From this and other evidence, paleontologists infer that the species always lived at low population density, and would probably have been noted by humans only as a minor element in any local fauna. Some giant deer bones had been found in apparent conjunction with human artifacts, but evidence remained inconclusive because the most undeniable criterion—representation of the species in Paleolithic art—had long yielded nothing positive. In a key article published in 1949, G. F. Mitchell and H. M. Parkes wrote: "It may perhaps be pointed out again that there are no representations of the Giant Deer in Paleolithic cave art."

Science tends to be difficult, subtle, ambiguous, and biased by all manner of social and psychic prejudice—though surely directed in a general way toward increasingly better understanding of a real world "out there." But every once in a while, we do achieve the reward of a simple, pristine, and undeniable fact—and then we can simply rejoice. In 1952, the first clear *Megaloceros* appeared on a newly discovered cave wall—a gift from our ancestors, and a positive solution to the question of whether humans ever interacted with giant deer. The cave of Cougnac, in south-central France, yielded three paintings of *Megaloceros,* two males and a female. Deer can be difficult to distinguish in cave art, for the paintings are partly symbolic and not entirely representational, and some species of deer differ only subtly. But the antlers of *Megaloceros* are so distinctive, and the second male of Cougnac so clearly painted, that the attribution could scarcely be doubted. Few deer have palmated, or platelike, antlers. The fallow deer, *Dama dama,* presents the only real possibility of confusion with *Megaloceros*—but the tines (points) of fallow deer spring from the posterior border of the palm, while giant deer tines project from the front edge. The Cougnac painting clearly depicts a large palmated antler, with tines springing from the anterior border.

One is a great discovery, but generalities require at least two. Forty-five years after the opening of Cougnac, giant deer remain rare in cave art—thus supporting my earlier inference that *Megaloceros* remained an uncommon

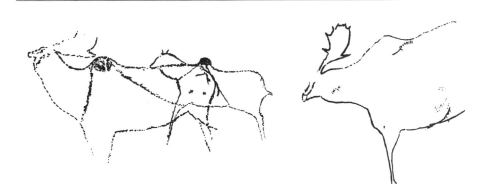

Megaloceros *from Cougnac Cave, southwest France.*

animal in ice-age Europe. Only four sites have been identified, and only one other locality satisfactorily affirms the Cougnac discovery. The cave of Peche Merle, known before Cougnac, contains a schematic figure of a probable giant deer—really little more than a rough finger sketch in clay. The antler does emit a strong whiff of *Megaloceros,* but I wouldn't put much money on this figure—and no one did before Cougnac confirmed the presence of giant deer in cave art. The recently discovered cave of Cosquer also contains two probable giant deer, but the identification requires comparison with Cougnac—and the Cosquer images alone could not have made a convincing case for *Megaloceros* in Paleolithic painting.

Thus, only one other find truly ranks as an independent affirmation of Cougnac, and as proof that our ancestors interacted with giant deer. The recently discovered cave of Chauvet (see the previous essay) contains two beautifully rendered giant deer. Neither painting includes the distinctive antlers, but all other defining features, as known from fossil bones and the Cougnac paintings, have been faithfully depicted, and the identification seems firm. (Both deer have been interpreted as females, and probably correctly so, but I wonder if one or both might not be males with shed antlers, for a small projection next to the ears could represent pedicels—the projecting bases of shed antlers, found only in males. In any case, the depiction of one sex alone makes sense in natural history, though such a decision may, of course, only reflect the symbolic purposes of the artists. In two excellent articles in the 1980s, paleontologist Anthony Barnosky proved from Irish localities that males and females lived in separate herds for part of the year—as many species of deer do today.)

For paleontologists, cave art provides precious evidence far beyond a mere proof of interaction with ancient humans. Consider the chief frustration of the conventional fossil record: that we must rely upon the evidence of bones and other preservable hard parts. So much that is so vital for any understanding—shapes of soft parts, colors, sounds, behaviors—simply do not fossilize. Our science depends crucially on modes of inference (often dubious or even fanciful) from a paltry preserved record to the richness of nature's totality. Sometimes we can draw a reasonable conclusion: When we find a shark's tooth embedded in an ammonite's shell, we know something about the diet of ancient fishes. But often we just can't tell: I cannot, for example, even imagine how we might learn crucial details about the emergence of human language, since so many millennia passed between original invention and the codification of any written system that might be preserved in the geological record. (In this context, I confess to wry amusement, bordering on annoyance, at the success of "dinomation" exhibits in museums—robotic dinosaurs that gyrate and howl. Public fascination for these models centers upon the very features—colors, sounds, and soft body flaps and frills—that must remain entirely conjectural.)

Paleontologists therefore treasure the rare geological circumstances that permit an occasional preservation of soft parts. Much of our most crucial knowledge about life's history requires these precious "windows" upon the complete anatomy of ancient creatures. We would never have known the full range of the Cambrian explosion if the Burgess Shale had not preserved soft parts as well as shells—for many of these earliest animals grew no hard parts at all. We would never have identified *Archaeopteryx* as the first bird if the lithographic limestone of Solnhofen did not preserve feathers as well as bones.

All these "windows" exist as a result of rare geological conditions, usually involving rapid burial in fine-grained sediments lacking both oxygen and a bacterial biota poised to decompose anything soft and organic. (Entombment in amber produces the same effect.) Only one new mode has been added by life's own complexity—unfortunately rather late in time, and quite limited in range. Human artists rendered the soft parts (and sometimes even the colors) of the fauna of ice-age Europe—and we, their descendants, are forever in their debt for this unique style of window into the past.

Even without a boost from cave art, we can learn more about ice-age mammals than about most other creatures of a more distant past—for complete and well-preserved skeletons can often be found, and the animals tend to be familiar as a consequence of their close affinity with living species. But many important features must still remain obscure when bones provide our only evidence. For example, we can infer the existence of an elephant's or tapir's trunk from the distinctive form of nasal bones, but we cannot learn much about size, color, or function. Similarly, fossil camels are almost always drawn without humps—not because we have any reason to assert their absence, but because we cannot infer their probable presence from bones alone.

Consider the most pressing question that bones alone cannot resolve about the form of giant deer. Our interest in this species has always focused—quite understandably, given their outlandish size—upon the antlers. How could an animal with a five-pound skull grow up to one hundred pounds of antlers year after year? Any structure of such size and exaggeration must require a substantial set of compensatory adaptations in other bodily parts—and much of the scientific discussion about giant deer has centered upon the redesign of the rest of the body to support the gigantic antlers. Adrian Lister argues, for example, that the remarkably thickened lower jaw bones may act as a source of recruitable storage—as they do to a lesser extent in some modern species of deer—for calcium that can be transferred to the antlers. Valerius Geist has shown that such enormous antlers impose severe constraints upon forage, for only a few plant species can supply enough minerals in the time required. Geist inferred that only willows could suffice, and he then discovered willow remains stuck in the teeth of giant deer fossils!

But most compensatory adaptations perform the more basic function of supporting the antlers in a biomechanical sense. For example, the top of the skull is unusually thick, and the first few vertebrae of the neck remarkably powerful and especially wide (as needed for the insertion of large muscles and ligaments that support the head). Most remarkably, as shown in the figure drawn by the great British anatomist Richard Owen (see page 191), the spines of the first few dorsal vertebrae (in the shoulder region) project far up from the backbone. In his *History of British Fossil Mammals and Birds,* published in 1846, Owen first noted the existence and significance of these projections as compensatory adaptations for the enormous antlers:

> The dorsal vertebrae are thirteen in number, and the anterior ones are remarkable for the length of the spinous processes which give attachment to the elastic ligaments supporting the head: those of the third, fourth, and fifth dorsals rise to a foot in height.

Modern studies have affirmed Owen's insight about the crucial status of spines on the dorsal vertebrae. A key structure, appropriately named the *ligamentum nuchae* (or neck ligament), supports the neck in modern deer (and other mammals) by attaching both to the occiput (back) of the skull and the first few cervical (neck) vertebrae in the head region, and then extending back to an insertion on the spines of the dorsal vertebrae in the shoulder region. The stronger the neck vertebrae, and the longer the spines of the dorsal vertebrae, the more powerful the ligament—all the better to hold up the massive head. All these structures, unsurprisingly, grow especially large in giant deer! (On the form and function of the *ligamentum nuchae,* see the 1985 article, listed in the bibliography, by N. J. Dimery, R. McN. Alexander, and K. A. Deyst.)

The long dorsal spines of giant deer imply some expression in the shape of the body—but of what form, and to what significance for the animal's function and behavior? Other large deer and related animals also possess elongated dorsal spines, and a broadly raised area around the shoulder region as a consequence—as in modern moose or bison.

But how could the even longer dorsal spines of giant deer influence the body's external form? Classic reconstructions of giant deer have either ignored this issue entirely and drawn a straight back (clearly an error), or they have depicted a long and low bulge on the back, as in modern moose.

For example, the "standard" painting of Charles R. Knight, the finest and most influential artist of prehistoric life, uses the "moose model" and draws an extensive but quite indistinct bulge.

At this point we are completely stuck; bones alone can take us no farther. And we would have been stuck forever—if not for a crucial gift from our artistic ancestors. The giant deer of Cougnac and Chauvet—two males and a female at Cougnac, and two probable females at Chauvet—teach us many things that we could not have known from standard paleontological evidence. Faces are slender and pointed as in most deer, not wide and fleshy as in moose. The males have broad and powerful necks. As in most deer, *Megaloceros* held its head low and in line with the backbone behind, not elevated above the back (with the neck virtually at right angles to the body) as in many reconstructions, including Owen's of 1846, as reproduced here.

But one feature of all these paintings stands out as a wonderful surprise and an unexpected result—and all commentators in the literature have emphasized this discovery. Paleolithic artists drew every giant deer, both male and female, with a large, discrete, localized, and prominent hump— not just a diffuse and indistinct raised area as minimally implied by the dorsal spines. (This unique and distinctive hump, first revealed at Cougnac, has since become the criterion for identifying giant deer in cave art. The poorly preserved figures at Cosquer can only be recognized by the hump. The two excellent paintings of Chauvet could be identified by other features but, in the absence of antlers for these probable females, the hump clearly marks the species.)

Moreover, the hump featured distinctive colors and markings that define the animal's appearance (and could also never be known from the conventional data of fossils). In all paintings that depict interior markings, the hump has been darkly colored—as a striking black blob, discretely outlined and covering the entire hump, in two animals from Cougnac; and as a more diffuse, but equally extensive, black patch on the first Chauvet female, and a thickened and accentuated border on the second animal. (Paleolithic artists drew the large male of Cougnac only in outline, with no interior

Skeleton of an Irish elk, as reconstructed by Richard Owen, 1846.

markings—presumably as a stylistic decision, not a representation. I do, of course, recognize that common features of these paintings may represent artistic convention or at least accentuation or exaggeration, rather than nature's measurable reality. But why draw such a discrete hump if none existed—especially since Paleolithic artists did accurately represent bison, elk, and reindeer with the broad and indistinct raised areas that usually cover the dorsal spines in large herbivores?)

The hump also serves as a focal point for other markings of uncertain meaning—probably either lines of color, folds of skin, or separations between regions of differing colors, lengths, or conformations of coat hair. All four drawings with interior features show a prominent line extending from the hump in a grand diagonal across the entire flank to the back legs. In three of four animals, another line runs from the forward edge of the hump on an opposite diagonal to the chest just above the forelegs. Finally, one animal from Chauvet bears a large diagonal black band across the entire neck, while a male from Cougnac carries a line of color in the same position.

The hump does form on the region of the back that overlies the dorsal spines. This "preexisting condition" of long spines to anchor a ligament for support of the heavily antlered head surely provides a substrate for the evolution of a discrete hump. But, equally surely, the hump of the giant deer has become more than a passive expression of the skeleton beneath—for passive expression only requires the broad and indistinct raised area developed by many other species over their enlarged dorsal spines. The hump, with its discrete and exaggerated form, and its bold accent by color (and by diagonal lines radiating out in both directions), must represent a distinctive product of evolution. But why?

We cannot answer this question with any confidence, but comparison with living relatives suggests a primary function in the signaling and display that accompany the central Darwinian activity of reproductive competition. In my earlier work, I suggested that giant deer might not have used their antlers in actual combat, but rather to announce status and to permit competition by symbolic posturing rather than by overt and harmful fighting. In short, and a bit crudely, bigger antlers win the bluff and secure more copulations for their bearers.

I now believe that I was wrong in making this suggestion. Later work by Tim Clutton-Brock and Andrew Kitchener has convinced me that giant deer almost surely used their antlers in actual combat. But fights for reproductive success among male deer (and other large mammals) also involve substantial ritual, posturing, and display—undoubtedly employing the antlers as a major element of the routine. Deer posture, bellow, and strut before an actual engagement. They often perform a "parallel walk" to observe the full length and conformation of a rival—and any feature that might accentuate an impression of power, fierceness, or bulk should help to establish a position of dominance. I agree with both Kitchener and Lister that the large and boldly colored hump would work especially well as an intimidating device and a mark of potential power. Kitchener writes that "the Irish elk [giant deer] probably assessed its opponents in a parallel walk which would have emphasized the massive shoulder hump and, hence, indicated body size and potential fighting ability." Lister adds that "the prominent, dark-colored dorsal hump would have formed part of the display *gestalt*."

I doubt that display for fighting can explicate the full function of the giant deer's hump—if only because females seem to grow an equally prominent hump, but presumably did not fight. Incidentally, the female hump also indicates a distinct functionality for this feature beyond the mere expression of underlying dorsal spines for the dorsal spines of non-antlered females are much shorter than those of males, but the female hump seems no smaller than the male version! Similar form and strength in both sexes may suggest further function as a general signal for recognizing other members of the species, or for other purposes yet unknown and unsuspected.

In any case, the uniquely shaped and distinctively functioning hump of the giant deer provides a superb illustration for a fundamental principle in evolutionary theory. The hump, we must presume, did not initially arise "for" eventual functions of display and recognition. The original structure probably developed, at first, as a simple and passive consequence of the underlying dorsal spines, themselves evolved for the very different, and obviously crucial, function of holding up a head with maximally heavy antlers—for all large mammals with dorsal spines must (and do) develop a broadly raised, if indistinct, bulge on the back over the spines. But most mammals never alter this indistinct structure in any substantial way; the bulging back remains a passive consequence of the underlying spines and may serve no function by itself at all. However, for some unknown reason, giant deer actively evolved this preexisting passive consequence into a discrete and prominent hump with complex adaptive functions in its own right.

Therefore, structures that initially arise as nonadaptive side effects of a primary adaptation (the raised area on the back as a necessary expression of underlying dorsal spines) may later be "coopted" for a special role vital to the animal's evolutionary success. Much of the fascination, the quirkiness, and the unpredictability of evolution lie in this principle of cooptation of structures initially evolved for other purposes, or for no purpose at all. Feathers that evolve as thermoregulatory devices in small running dinosaurs get coopted for flight in birds. Brains that evolve for whatever our australopithecine ancestors needed on the African savannahs get coopted by later Cro-Magnons for artistic expression and utility—and so (and only for this reason) do we learn about another coopted structure: the hump of the giant deer!

The hump of *Megaloceros* may intrigue us as an illustration of this important general principle. Nonetheless, I do not wish to advance such an argument as our major reason for fascination with this bump on a back— however dear the principle of cooptation may be to me (as a prominent theme in my own technical writings), or however much the movement from little fact to large generality forms the trademark of these essays. After all, the principle of cooptation has been both well established and well illustrated for many years. To be sure, another good example never hurts— especially for such a striking structure in such a fascinating species. But we break no theoretical ground with this illustration.

May I suggest instead that we should value the hump of the giant deer primarily for a very different reason—as an item of natural history, precious beyond words simply because it once existed, and because we would never have known either its factuality or its fascination if our ancestors had not been moved to leave such beautiful visual records.

For simple items regulated by natural laws, we can often infer existence (and conformation) without actual observation. We know the record of solar eclipses for the last several thousand years, even though many went unrecorded in all human chronicles. But for complex items of natural history, unrepeatable in their unique and detailed glory, and crucially dependent upon a contingent and unpredictable sequence of prior historical states, we cannot know their existence unless the paltry and grossly imperfect records of history leave direct evidence. Every item of natural history is both a joy to behold and an instrument for our potential enlightenment. But the vast majority of items have been permanently lost in the bottomless pit of history's failure to record. And a loss at any moment is a loss forever.

You may say, "So what, we are surrounded by such a plethora of items, and we can't know everything." But I am insatiably greedy and infinitely curious. Each and every loss becomes an instance of ultimate tragedy—something that once was, but shall never be known to us. The hump of the giant deer—as a nonfossilizable item of soft anatomy—should have fallen into the maw of erased history. But our ancestors provided a wondrous rescue, and we should rejoice mightily. Every new item can instruct us; every unexpected object possesses beauty for its own sake; every rescue from history's great shredding machine is—and I don't know how else to say this—a holy act of salvation for a bit of totality.

We will never know the Paleolithic painter who rescued this precious fact—the hump of the giant deer—and vouchsafed it to us, his grateful descendants. To this anonymous person, I can only say: "You're a better man than I am . . ." For I can only report and interpret, but you salvaged a real and true item of earthly beauty

1 0

OUR UNUSUAL UNITY

A TRULY STUPID MISTAKE OFTEN INITIATES A PATH TO ENLIGHTENMENT. Fortunately, my latest experience of this common phenomenon occurred (and got corrected) in total privacy—so I can avoid embarrassment because no one need know! *The Herald,* Zimbabwe's major newspaper, printed a government notice in its issue for January 14, 1997: *Licensing of dogs and cycles.* The annual fee, they reported, would be twenty Zimbabwean dollars (about $2.00 U.S.) for a bicycle, and thirty for a tricycle. I laughed to myself at the blatant absurdity of charging more for a kid's toy than an adult's necessity—an amusement no doubt tinged (I must admit) by residual and unconscious racism so pervasive in our culture that even white folks of decent will cannot entirely extinguish the blight: those primitive Africans got it backward again. But the joke was entirely on me—for I soon remembered that local tricycles are the three-wheeled, human-powered vehicles that serve as short-haul taxis, or for transporting heavy goods, in so much of the non-Western world. These sturdy adult tricycles are larger and generate more income than a bicycle—and may therefore be fairly taxed at a higher rate.

This example of one of the most common fallacies in human reasoning—the elevation to universal status of a local, limited, and potentially false belief held by an individual or a culture—did no harm and lasted only a few minutes. But other more potent and pervasive cases often serve as the greatest impediments to improved scientific and scholarly understanding. The larger theme behind my little blunder about two- and three-wheeled vehicles—the assumption that human history should progress in a linear sequence of improvement (with Africans behind Europeans)—may be the most harmful and widespread of all culturally embedded errors falsely promoted to universal truth.

I recently encountered a striking example during some heavy travel between two monthly essays. I saw, for the first time, the magnificent remains of the great Mayan cities of Chichen Itza and Uxmal. Among the many anomalies presented by this ancient and complex Mesoamerican civilization, the problem of deciphering Mayan writing stands out. Mayan culture peaked in the second half of our first millennium and then mysteriously collapsed from the ninth to the tenth century A.D. (Several resuscitations occurred thereafter, partly in amalgamation with other Mesoamerican groups—and Mayan people, speaking Mayan languages, still inhabit Guatemala, the Yucatán, and surrounding areas. But knowledge of the classical writing system, and much of their elaborate astronomical and calculational learning, did not survive the European invasions.)

Spanish conquerors destroyed most of the Mayan books (written on paper made from bark, and folded accordion-style)—and only four codices survive. But Mayan writing appears on hundreds of large stone stelae, originally erected as ceremonial proclamations in front of major buildings, and as numerous inscriptions on walls, statues, and pots. The recent deciphering of this script—a mixture of symbols for syllables and for entire words (and therefore similar in concept to Egyptian hieroglyphs, though independently invented)—ranks as one of the greatest scholarly achievements of the twentieth century.

We may rejoice in this success, and in the striking reinterpretation of Mayan history thus provided, but we must also wonder what delayed the decipherment for so long—for as Michael D. Coe points out in his recent and justly acclaimed book, *Breaking the Maya Code,* the tools for successful resolution, and adequate data for the task, had been potentially available from the beginning of serious Mayan scholarship in the mid-nineteenth century.

The reasons are complex and many, including the cruel and systematic destruction of Mayan documents by early Spanish colonialists, but Coe shows that the old error of construing human history as linear progress also played a major role. Since the Mayans peaked so long ago (while Europe remained a backwater), and belonged to an ethnic group judged inferior by many scholars of European extraction, several leading experts on Mayan culture simply refused to believe that these inscriptions could represent a complete written language. Mayan writing must, they argued, represent the crude pictorial scribbles of limited people who, despite surprising and considerable achievements in architecture and astronomy, could never master the full complexity of recorded language.

For example, Coe quotes one Mayan scholar who, in 1935, denounced the good start made by Benjamin Lee Whorf, a great linguist of the last generation. Whorf had correctly ascribed phonetic value to the Mayan glyphs, but his critic replied that Mayan symbols could only represent "embryo" writing—crude pictures with limited informational content, and not full sentences with grammar. Invoking this progressivist, linear (and racist) assumption, Whorf's critic wrote:

> E. B. Tylor said long ago that writing marked the difference between civilization and barbarism . . . The fact remains that no native race in America possessed a complete writing and therefore none had attained civilization according to Tylor's definition.

Coe also shows how the "hyper-evolutionism" of Sylvanus Morley, the dominant Mayan scholar in the first half of our century, also became an impediment to deciphering this largely phonetic script with complete grammar. Coe writes:

> Sylvanus Morley . . . proposed that writing systems had progressed from pictographic, through ideographic (with Chinese given as an ideographic system *par excellence,* since according to Morley each sign stands for an idea), to phonetic.

Since Morley viewed the Mayans as more primitive than the Chinese—and therefore largely in the pictographic stage—he could never have deciphered their predominantly phonetic writing!

Let me now crank up the scale for this cardinal error of linearization one notch further—from my personal mistake about a modern nation, to a serious blunder that long delayed the explanation of an entire culture with an extended history, to a major misconception that often stymies our understanding of human evolution as a totality.

We may legitimately speak of "general trends" in human evolution. We can also scarcely doubt that increasing brain size represents both a major trend and the key to our species's extraordinary history of spread and domination. Such a statement does not, however, necessarily imply that human history—from the split, 6 to 8 million years ago, of our ancestors from the common stock that also generated our closest cousins (chimps and gorillas), to our current exalted state—should be interpreted as a linear series of advancing steps in brain power, with any stragglers, or groups that failed "to go with the program," relegated to extinction as side branches in an inevitable cul-de-sac, or dead end.

Many paths and mechanisms can lead from a small-brained beginning to a top-heavy current status. To cite the most radical evolutionary alternative to the traditional linear view—a false extreme, to be sure, but providing as much partial insight as the equally erroneous linear alternative—suppose that an ancestral Species A, with an average brain volume of 300 cubic centimeters, generated five new species, all during a short and crucial period, say between 2.2 and 2.0 million years ago. These five species arise with different average brain volumes—B at 500 cc, C at 700, D at 900, E at 1,100, and F at 1,300—and do not alter these figures during their geological lifetimes. All six species (A and the five descendants) live for 2 million years with no further change. (They may never even come into direct competition, for each may inhabit a different continent—the result of A's rapid spread around the world and equally quick evolution to B, C, D, E, and F in five separate areas.) Finally, species A through E become extinct and only F survives. We call F *Homo sapiens*.

In both extreme cases, human ancestors begin at 300 cc, and peak today at 1,300 cc. Both schemes invoke a metaphor of struggle and persistence—the slow climb up a ladder in the traditional linear view, and success in hanging on through every adversity in the "bush pruning" alternative. Both views are also clearly wrong in their exclusive versions. Why, then, do we tend to feel comfort and affinity for the linear scheme, while regarding the "bush pruning" alternative as laughable and inexplicable nonsense, no doubt introduced by yours truly to satisfy some perverse and personal whim.

Yet I wish to argue (1) that both views express important partial truths; (2) that we have favored the linear view primarily in obedience to the disabling cultural bias illustrated by the earlier examples in this essay; (3) that the history of twentieth-century ideas about human evolution can be epitomized by growing strength of the "bush making and pruning" view—and the retreat of the linear view—all leading to a proper balance; and (4) that a new discovery, announced in December 1996 (and inspiring this essay), provides strong and unexpected support for bushiness as the usual condition of the human lineage. (I shall, for the rest of this essay, refer to the two modes of thought as "linear" and "bushy" accounts of evolutionary trends.)

My friend and closest colleague, Niles Eldredge, has labeled these two approaches to trends as "taxic" and "transformational"—or "based on the production of many separate species" (formally named groups of organisms, such as species and genera, are called "taxa"), versus "propelled by the advantages of certain traits" (big brains, for example) in competition among varying individuals within a single group. The two views differ most significantly in their primary "motors" for generating trends. In the bushy, or taxic, theory, trends require a substantial production of independent species, for net change in a lineage depends upon a differential survival and further proliferation of some species versus the extinction of others. In the linear, or transformational, theory, trends require no bush of species, but arise by the competitive success of favorable traits in a gradually progressing unit. (Of course, supporters of the linear view do not deny that lineages may also produce new species by branching, but these scientists tend to separate the progressive carrier of the trend from doomed side branches. In other words, for the linear transformationist, the production of numerous species does not contribute to the major progressive trends of life's history.) Ernst Mayr, the dean of American evolutionists, and a strong supporter of copious speciation as a central ingredient in evolutionary trends, expressed the contrast well by writing:

> I feel that it is the very process of creating so many species which leads to evolutionary progress. Species, in the sense of evolution, are quite comparable to mutations. They also are a necessity for evolutionary progress, even though only one of many mutations leads to a significant improvement of the genotype . . . Seen in this light, it appears then that a prodigious multiplication of species is a prerequisite for evolutionary progress . . . Without speciation, there would be no diversification of the organic world, no adaptive radiation, and very little evolutionary progress. The species, then, is the keystone of evolution.

Nonetheless, the linear view has, until recently, strongly dominated traditional thinking about human evolution. For example, the hoary and clichéd concept of a "missing link" presupposes linearity—for links are joining points in a sequence. Evolutionary bushes may be riddled with all the absences and uncertainties imposed by our poor fossil record, but a bush cannot feature a single and crucial "missing link."

Moreover, the linear view has not just been accepted passively or unthinkingly—as a simple expression of an unquestioned bias. The idea of coexistence among several hominid species has been actively denied, attacked, and even stigmatized as bad biological reasoning. For example, when I was a graduate student in the 1960s, an idea called "the single species hypothesis" still enjoyed strong, probably majority, support among students of human evolution. According to this theory, only one hominid species could, in principle, occupy a single region at any one time. Thus, since most of our evolutionary history had unfolded on the single continent of Africa, our trends must arise by linear transformation, with only one species living at any moment, slowly perfecting itself toward the next stage. Advocates cited (I would say mis-cited) the ecological principle that only one species can occupy any "niche"—or suitable environment for "making a living." Beetles have "narrow" niches, so several species can live in one area—some on bark, some on the ground, some high in trees. But hominids, with our unique invention of "culture" (however primitive at first), occupy such a "broad" niche that no single place can house more than one species.

The leading basic textbook in physical anthropology at the time (*Human Evolution,* by C. L. Brace and M. F. Ashley Montagu, 1977, first edition 1965), held that "the known fossils are most realistically placed in a linear evolutionary relationship." The authors specified four sequential stages—australopithecine, pithecanthropine, Neanderthal, and modern—and justified their sequence by the "single species hypothesis." They wrote:

> Culture as a major means of adaptation is unique in the world of living organisms, and for all important purposes can be considered an ecological niche in itself—the cultural ecological niche. There is an evolutionary principle based on the logic of efficiency which states that, in the long run, no two organisms can occupy the same

ecological niche. In the end, one will out-compete the other and retain sole possession of the niche in question. Applied to the primates, this should mean that no two forms could occupy the cultural ecological niche for any length of time.

C. Loring Brace, one of the authors of this text, has continued to resist the notion of bushiness in hominid evolution. In the 1991 edition of his popular text *The Stages of Human Evolution,* Brace acknowledges only one side branch in the entire history of human evolution—and he calls this substantial lineage of robust australopithecines a "twig"!

> My own view, however, is represented by the final unilinear arrangement, where the Australopithecines evolved into the Pithecanthropines which in turn evolved into the Neanderthals throughout the whole of the inhabited Old World, and these finally became transformed into the various modern populations alive today. I have left off the Australopithecine twig that became hyper-robust and died out . . . just to give a streamlined version of my general view.

Brace dismisses the idea that two (or more) human species might have interacted in one place. He even invents the label of "hominid catastrophism" to stigmatize the view (now favored by most paleontologists, particularly for the replacement of Neanderthals by moderns in Europe) that a temporal transition from one species to another might arise by immigration of the later species from another region (followed by local extinction of the original inhabitants), rather than by linear evolutionary transformation. Brace writes:

> The result is remarkably like the picture presented by Cuvier's catastrophism early in the nineteenth century which regards change as occurring suddenly, for undiscoverable reasons, and away from the region under examination. The new form, which spreads by migration, then prevails until the next sudden change.

Yet, of all alterations in thinking about human evolution that have occurred during my professional lifetime, none has been more transforming, or further ranging in implications, than the increasing documentation of substantial bushiness throughout most of hominid history. Our present reality of one worldwide species represents an oddity, not the norm—and we have been fooled by our bad habit of generalizing a transient and contingent present.

I would summarize this fundamental change from the linear to the bushy view of our evolutionary history in five chronological discoveries and arguments, with the latest news as the fifth finding

1. *TWO BRANCHES OF AUSTRALOPITHECINES.* When South African scientists described *Australopithecus,* the genus ancestral to our own *Homo,* in the 1920s, they designated two major branches or species, *Australopithecus africanus* and *A. robustus* (known in later literature as the gracile and robust forms). Thus, a bushy theory for our early days enjoyed some support from the start. But proponents of the single-species hypothesis either viewed the two names as improperly given to males and females of a single species, or (as in the quote from Brace previously cited) regarded the robust lineage as a doomed and insignificant side branch, probably driven to extinction by our superior forebears, the graciles.

However, in 1959, Mary Leakey found a key specimen with robust features so exaggerated that sexual variation within a single species became implausible as an explanation for the extent of difference. The probable coexistence of two australopithecine lines could no longer be denied—and the purest version of the single-species hypothesis died. (Mary Leakey originally called this skull *Zinjanthropus;* we now generally designate this form as a separate, so-called hyper-robust species, *Australopithecus boisei.*)

2. *COEXISTENCE OF AUSTRALOPITHECUS AND HOMO.* Linearists could still adopt a fallback position. They could brand the robust (and hyper-robust) australopithecines as an insignificant blind alley, regard the graciles as lin-

early ancestral to our own genus *Homo,* and then apply the single-species hypothesis to *Homo* alone, drawing a line from *Homo erectus* ("Java" and "Peking" man in the older texts) through Neanderthal to our current exaltation. But then, in the mid-1970s, Richard Leakey (Mary's son) found hyper-robust specimens in the same strata that yielded bones of African *Homo erectus* (sometimes called *Homo ergaster,* but little different from the Asian *Homo erectus* of Indonesia and China). No one could possibly encompass this range of variation within the boundary of a single species. If the most extreme of the robust australopithecines coexisted with the most advanced members of our own ancestry, then the old line of progress had become an undeniably diverging bush.

3. *THE PLETHORA OF AFRICAN SPECIES BETWEEN 3 AND 2 MILLION YEARS AGO.* Two branches destroy the linear theory, but don't build a very impressive bush. In the twenty years since Richard Leakey's discovery of these two irrefutably coexisting species, further research on hominid history has stressed one primary theme above all others: The bush gets bushier and bushier. To summarize a great deal of elegant research in too short a statement: We have no evidence for more than one species during the earliest period from 3.0 to more than 4 million years ago. (For most of this interval, we know only *Australopithecus afarensis,* the famous "Lucy" of our popular literature.) But between 3.0 and 2.0 million years ago (and mostly during the last half-million years of this interval), a virtual explosion of hominid species occurred, on both major branches of the hominid bush—that is, both within the ancestral genus *Australopithecus,* and within the derived genus *Homo.* The accompanying chart, presented in Donald Johanson and Blake Edgar's recent book *From Lucy to Language,* shows as many as six coexisting hominid species during this period, three within our own genus *Homo.*

4. *BUSHINESS IN LATER HUMAN HISTORY: THE NEANDERTHAL ISSUE.* Linear preferences die hard. I think that all major students of the subject now accept substantial bushiness, and coexistence of several species in Africa,

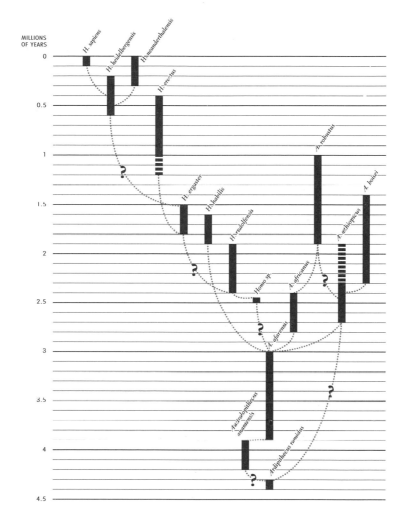

MILLIONS
OF YEARS

The "branching bush" of human evolutionary history: between 2.5 and 1.5 million years ago, as well as during the last half-million years, several hominid species coexisted.

during early hominid history, but a version of the old linear view still persists as a popular (though, I judge, dwindling) theory for later human history during the past million years or so, and especially for the origin of *Homo sapiens.* This debate has been prominently featured in the press (and treated in several of these essays) as a conflict between the "multiregional" and "out-of-Africa" theories for modern human origins. Multiregionalism will probably be remembered as the last post of the linear view. Under this model, all hominid evolution occurs in Africa (admittedly in a fairly bushy manner) until the origin of *Homo erectus*. This species then spreads out to all the Old World continents between 1.5 and 2 million years ago. The three major populations of *Homo erectus,* in Africa, Europe, and Asia, then evolve in parallel (abetted by a low level of migration and consequent mixing among the three groups) toward *Homo sapiens.* Such an idea represents linearity with a vengeance—as all subgroups within a single species move onward (and brainward) in the same optimal direction. In Europe, for example, *Homo erectus* evolves to Neanderthal, and Neanderthal transforms to *Homo sapiens*—one species at any time, but constantly on the upward move.

The out-of-Africa alternative may best be understood as a particular version of the bushy perspective. *Homo erectus* moves out to all three Old World continents. *Homo sapiens* arises as a branch (the bushy view) from one of these populations, not as a terminus to a universal trend. *Homo sapiens* then spreads as a second diaspora from its place of origin, presumably Africa on both genetic and paleontological grounds. But *Homo erectus* (or its descendants) already inhabit Europe and Asia—so African *Homo sapiens* arrives as a second human species (bushy coexistence again), and eventually supplants the original form. Under this bushy view, Neanderthal and modern *Homo sapiens* are separate (and potentially coexisting) human species, not the before and after of a single linear transformation—for Neanderthal branched from European *Homo erectus* (or its descendants), while forebears of modern Europeans arrived from Africa after a separate origin from African *Homo erectus* populations.

In my reading, and as summarized elsewhere (perhaps best in the recent book by C. Stringer and R. McKie, *African Exodus: The Origin of Modern Humanity*), the balance of recent evidence tilts strongly (perhaps conclusively) to the out-of-Africa view, and therefore to the predominance of bushiness over linearity as a central theme in human evolution. (Incidentally, this new and emerging consensus is the very view that Brace so scornfully rejected and labeled as "hominid catastrophism"—the idea that *Homo sapi-*

ens arrived from Africa as a second wave and supplanted Neanderthal, in contrast with the only reconstruction that Brace regarded as "evolutionary," that is, the linear passage of Neanderthal to modern humans. In fact, both views are equally consistent with an evolutionary perspective. The contingent and empirical data of actual history, not preferences of theory [laden with a complex range of unconscious biases], must decide the issue.)

5. *MORE BUSHINESS IN LATER HUMAN HISTORY: NEW DATA FROM ASIA.* If Neanderthal and *Homo sapiens* coexisted as independent species in Europe, thus refuting the linear view, what happened in eastern Asia, where Dubois first discovered *Homo erectus* in the 1890s, and where this ancestral species enjoyed long and widespread success? In the multiregional view, these Asian *Homo erectus* populations evolved directly into modern Asian groups of *Homo sapiens*. In the bushy alternative, *Homo sapiens* arrived (ultimately from Africa) as a second wave of migration, and may have coexisted for a time with Asian *Homo erectus* or its descendants. The obvious test between these starkly different views requires a fossil record, either of intermediacy or of coexistence, during the crucial time of transition between the two species. But such decisive data have not been available, because the youngest known Asian *Homo erectus* (from China) range from about 290,000 to 420,000 years old, while the oldest Asian *Homo sapiens* specimens are only about 40,000 years old. Thus, we had no evidence at all for the crucial intervening years.

Eugen Dubois first discovered *Homo erectus* in Java during the early 1890s—and these specimens, from Trinil, remain the most famous Indonesian representatives of the species. But, in the early 1930s, Dutch geologists discovered a suite of twelve hominid calvaria (skull tops lacking the facial skeleton and upper jaws) from the nearby site of Ngandong on the banks of the Solo River. These specimens—variously known in the older literature as "Solo man" or *"Homo soloensis"*—have engendered a long and substantial debate about their identity, but a present consensus considers them as members of Dubois's species, *Homo erectus.*

Yet, while anthropologists had finally reached some agreement about their identity, the age of the Solo specimens remained unknown. This crucial issue may now have been resolved—and in a surprising manner—by an article that appeared in the December 13, 1996, issue of *Science* magazine: "Latest *Homo erectus* of Java: Potential contemporaneity with *Homo sapiens* in Southeast Asia," by C. C. Swisher III, W. J. Rink, S. C. Anton, H. P. Schwarcz, G. H. Curtis, A. Suprijo, and Widiasmoro (yes, the last author's name is complete; most Indonesians, like former leader Suharto, or past boss Sukarno, use only one name). The curators of the Solo calvaria would not let these authors use the original material for dating (since the methods destroy parts of the specimens). So Swisher and colleagues collected bovid teeth (cattle and their relatives) from two sites in the same stratum that yielded the hominid calvaria. They applied two independent techniques of radiometric dating, and reached the same surprising conclusion—ever so gratifying for fans of the bush—that the Solo hominid specimens lived between 27,000 and 53,000 years ago. If these conclusions stand up to later scrutiny, then *Homo erectus* did not transform to modern humans in Asia—for the two species coexisted as independent entities about forty thousand years ago.

Moreover, and moving to the general statement that inspired this essay, if we now consider the whole earth at forty thousand years ago, we note a bush of three coexisting human species—*Homo neanderthalensis* in Europe, surviving *Homo erectus* in Asia, and *Homo sapiens* continuing a relentless spread throughout the habitable world. This collection of three might not match the richness of an African bush of some half a dozen species about 2 million years ago, but the conclusion that three human species still coexisted as recently as thirty to forty thousand years ago does require a major reassessment of conventional thinking. Our modern world represents the oddity, not the generality. Only one human species now inhabits this planet, but most of hominid history featured a multiplicity, not a unity.

I have focused this essay upon one of the great unconscious biases—our persisting preference for viewing history as a tale of linear progress—that so often stymie our interpretations of evolution and the history of life on Earth. But we should also recognize this other, rather more "homey" or obvious bias—our tendency to view a comfortable and well-known current situation as a generality rather than a potential exception. Such an attitude also has a highfalutin name in the history of science—*uniformitarianism,* or using the present as a key to the past.

As many scholars have pointed out (including yours truly in his very first published paper of 1965), uniformitarianism is a complex term with multiple meanings, some legitimate, but some potentially false and surely constraining. If we only mean that we will regard nature's laws as invariant in space and time, then we are simply articulating a general assumption and rule of reasoning in science. But if we falsely extend such a claim to current *phenomena* (rather than universal *laws*)—and argue, for example, that continents must always be separated because oceans now divide our major landmasses, or that mass extinction by meteoritic bombardment cannot occur because we have never witnessed such an event during the short span of recorded human history—then we surely go too far. The present range of observed causes and phenomena need not exhaust the realm of past possibilities.

In this case, we shall be massively and seriously fooled if we extrapolate a current reality to a general situation in the history of human evolution. Most of hominid history has featured a bush, sometimes quite substantial, of coexisting species. The current status of humanity as a single species, maximally spread over an entire planet, is distinctly odd.

But if modern times are out of joint, why not make the most of it? I last visited Africa more than ten years ago, and that voyage (to lecture in Nairobi and do some fieldwork with Richard Leakey at Lake Turkana) led to some musings that culminated in an essay titled "Human Equality Is a Contingent Fact of History." I argued that the happenstance of a surprisingly recent common ancestry for all modern humans had made our so-called races effectively equal in biological capacities (while individuals within all groups differ widely, of course).

I couldn't help revisiting this theme as I surveyed many sites of human hope, disappointment, and struggle in Kenya, Malawi, and Zimbabwe. (I went to Africa in my role as a trustee of the Rockefeller Foundation, and we visited many of their social, medical, and agricultural projects—including a clinic for treating sexually transmitted diseases among prostitutes in one of the worst slums in Nairobi, and a series of schemes to improve corn yields in desperately poor agricultural villages of Malawi.)

In the most memorable event of this trip, we spent an entire morning talking with the women farmers of a small Malawian village. This ample time gave us leisure to explore in depth, and to listen and observe with great care. My mind wandered over many subjects, but I kept returning to a single theme. I could not imagine a greater difference between earthly communities—a senior American Ivy League professor, and an illiterate Malawian farmer, twenty-five years old, with five children (the oldest already eleven), and an annual family income of about eighty dollars. Yet her laughter, her facial expressions, her gestures, her hopes, her fears, her dreams, her passions, are no different from mine. One can understand the argument for human unity in a purely intellectual and scientific sense, but until this knowledge can be fleshed out with visceral experience, one cannot truly know in the deeper sense of compassion.

If our current times are peculiar in substituting the bushy richness of most human history with an unusual biological unity to undergird our fascinating cultural diversity, why not take advantage of this gift? We didn't even have such an option during most of our tenure on Earth, but now we do. Why, then, have we more often failed than succeeded in the major salutary opportunity offered by our biological unity? We could do it; we really could. Why not try sistership; why not brotherhood?

IV

OF HISTORY AND
TOLERATION

11

A CERION FOR CHRISTOPHER

IF CHINA HAD PROMOTED, RATHER THAN INTENTIONALLY SUPPRESSED, THE technology of oceanic transport and navigation, the cardinal theme for the second half of our millennium might well have been eastward, rather than westward, expansion into the New World. We can only speculate about the enormously different consequences of such an alternative but unrealized history. Would Asian mariners have followed a path of conquest in the Western sense? Would their closer ethnic tie to Native Americans (who had migrated from Asia) have made any difference in treatment and relationship? At the very least, I suppose, any modern author of a book printed on the American East Coast would be writing this chapter either in a Native American tongue or in some derivative of Mandarin.

But China did not move east, so Christopher Columbus sailed west, greedy to find the gold of Cathay and the courts of the grand Khan as described by his countryman Marco Polo, who had traveled by different

215

means and from the other direction. And Columbus encountered an entire world in between, blocking his way.

I can think of no other historical episode more portentous, or more replete with both glory and horror, than the Western conquest of America. Since we can neither undo an event of such magnitude nor hope for any simple explanation as an ineluctable consequence of nature's laws, we can only chronicle the events as they occurred, search for patterns, and seek understanding. When dense narrative of this sort becomes a primary method of analysis, detail assumes unusual importance. The symbolic beginning must therefore elicit special attention and fascination. Let us therefore take up an old and unresolved issue: Where did Columbus unite the hemispheres on October 12, 1492?

Surrounded by hints of nearby land, yet faced with a crew on the verge of rebellion, Columbus knew that he must soon succeed or turn back. Then, at 2:00 A.M. on the morning of October 12, the *Pinta*'s lookout, Rodrigo de Triana, saw a white cliff in the moonlight and shouted the transforming words of human history: *"Tierra! Tierra!"*—land, land. But what land did Columbus first see and explore?

Why should such a question pose any great difficulty? Why not just examine Columbus's log, trace his route, look for artifacts, or consult the records of people first encountered? For a set of reasons, both particular and general, none of these evident paths yields an unambiguous answer. We know that Columbus landed somewhere in the Bahama Islands, or in the neighboring Turks and Caicos. We also know that the local Taino people called this first landfall Guanahaní—and that Columbus, kneeling in thanks and staking his claim for the monarchs of Spain, renamed the island San Salvador, or Holy Savior. But the Bahamas include more than seven hundred islands, and several offer suitable harbors for Columbus's vessels. Where did he first land?

Navigation, in Columbus's time, was far too imprecise an art to provide much help (and Columbus had vastly underestimated the earth's diameter, thereby permitting himself to believe that he had sailed all the way to Asia).

A *Cerion* for Christopher

Mariners of the fifteenth century could not determine longitude, and therefore could not locate themselves at sea with pinpoint accuracy. Columbus used the two primary methods then available. Latitude could be determined (though only with difficulty on a moving ship) by sighting the altitude of Polaris (the North Star), or of the sun at midday. A ship could therefore sail to a determined latitude and then proceed either due east or west, as desired. (Columbus, in fact, was a poor celestial navigator, and made little use of latitudes. In one famous incident, he misidentified his position by nearly twenty degrees because he mistook another star for Polaris.)

In the other time-tested method, called dead reckoning, one simply takes a compass bearing, keeps track of time, judges the ship's speed, and then plots the distance and direction covered. Needless to say, dead reckoning cannot be very precise—especially when winds and currents complicate any determination of speed, and when (as on Columbus's ships) sailors measure time by turning a sandglass every half hour! Columbus, by all accounts, was an unusually skilled and spectacularly successful dead reckoner, but the method still doesn't allow any precise reconstruction of his routing.

We are further hindered by a paucity of documents. Columbus's original log, presented to Queen Isabella, has been lost. A copy, given to Columbus before his second voyage, has also disappeared. Bartolomé de Las Casas, the Dominican priest who spoke so eloquently for the lost cause of kindness, made a copy of Columbus's second version—and our modern knowledge derives from this document. Thus, we are using a copy of a copy as our "primary" text, and uncertainties therefore prevail on all crucial points.

The best possible source of evidence—artifacts and recorded histories kept by an unbroken line of original inhabitants—does not exist for another reason that motivated this essay. In the first case of New World genocide perpetrated by the Old, Spanish conquerors completely wiped out the native Bahamians within twenty years of contact, despite (or rather, one must sadly say, enabled by) the warm and trusting hospitality shown to Columbus by the peaceful Tainos.

With so little data to constrain speculation, virtually all major Bahamian islands have been proposed as San Salvador, the site of Columbus's first landfall. (The Turks and Caicos Islands, just to the southeast of the Bahamas, form a politically separate entity. They are, however, geographically and ecologically continuous with the Bahamas, and therefore figure in this discussion as well.) The major contenders include Watling Island, Cat Island, Mayaguana, Samana Cay, Grand Turk, and several of the Caicos. Cat Island held an early advantage, and once even bore the name San Salvador in acknowledgment. But, in 1926, the Bahamian government, persuaded by a growing consensus, transferred Columbus's designation to Watling Island—the favored site, and Columbus's name bearer ever since.

Two traditional sources of evidence favor Watling as San Salvador: correspondence of size and topography with the fairly detailed descriptions of Columbus's log (as known by the Las Casas copy), and nautical tracing of Columbus's route for the rest of his first voyage, from San Salvador to other Bahamian islands, and finally to Cuba and Hispaniola. Samuel Eliot Morison's "semiofficial" case, made in his 1942 classic, *Admiral of the Ocean Sea,* remains the standard expression of this favored hypothesis.

During the past twenty years, archaeology has provided a third source of evidence from excavations made by Charles A. Hoffman and others at Long Bay, within sight of the favored location for Columbus's landfall. (A good account of this work, and of virtually all else connected with the discussion of Columbus's initial landing, may be found in the Proceedings of the 1986 San Salvador Conference on Columbus and His World, edited by Donald T. Gerace and published by the Bahamian Field Station on San Salvador.) Along with native pottery and other Taino artifacts, several European objects were found, all consistent with Spanish manufacture at the right time, and all eminently plausible as items for trade—glass beads, metal buckles, hooks, and nails. One discovery exceeded all others in importance: a single Spanish coin of low value, known as a *blanca*—the standard "small change" of the times, and surely the most common coin in circulation among Columbus's men. Moreover, this particular *blanca* was only issued between 1471 and 1474, and no comparable, copper-based coin was minted again until 1497.

Of course, these finds do not positively identify San Salvador as the first landfall for two reasons: Columbus visited several other Bahamian islands on this voyage, and the local Tainos moved freely among adjacent islands. In fact, three days after his first landing, and again on open waters, Columbus encountered a Taino in a canoe, carrying some beads and *blancas* received in trade on San Salvador.

Nonetheless, and all other things considered, the archaeological evidence supports the usual view that San Salvador has now been correctly identified. Still, everything cited so far relies upon European impressions or artifacts. Wouldn't we welcome some hard data from the other side for corroboration? How about one distinctive item of local history, either natural or cultural?

I do not mean to exaggerate the current uncertainty in this debate. Most experts seem satisfied that Columbus first landed on the island now called San Salvador by the Bahamian government. Nonetheless, several dogged

Erection of monuments at Columbus's reputed landing sites has been something of a cottage industry on San Salvador for more than a century. The author stands at each of the three markers that adorn the island: the Chicago Herald's *at Crab Cay (left), the Japanese contribution (center), and the "official" cross at Long Bay (right).*

and knowledgeable opponents still advocate their alternatives with gusto, and the issue remains vigorously open. I recently spent a week on San Salvador, where I sifted through all the evidence and visited all the sites. I found no reason for dissatisfaction with the conventional view. Nonetheless, if only because we prefer near certainty to high probability, I write this essay to announce that I could truly resolve any remaining doubt about Columbus's first landfall if only the good admiral had added one little activity to the usual drill of kissing the ground, praising God, raising the flag, claiming sovereignty, and trading with the locals. If Christopher Columbus had only picked up (and properly labeled, of course) a single shell of my favorite animal, the land snail *Cerion*—and they are so common that he was probably kneeling on one anyway!—I would know for sure where he had landed.

No one can be objective about his own children, but *Cerion* truly ranks as a natural marvel, and an exemplar of evolution for a particular reason well illustrated by its potential utility for identifying San Salvador. In shell form, *Cerion* may be the most protean land snail in the world—and evolutionists thrive on variation, the result and raw material of biological change. *Cerion* ranges in size from dwarfs of 5 millimeters to giants more than 70 millimeters in length (for folks wedded to good old ways, 25.4 millimeters make an inch)—and in shape from pencil-thin cylinders to golf balls.

Naturalists have named more than six hundred species from *Cerion*'s two major geographic centers in Cuba and the Bahama Islands. Most of these names are technically invalid because members of the respective populations can interbreed, but the designations do record a striking biological reality—that so many local populations of *Cerion* have evolved unique and clearly recognizable shell forms. In particular, nearly every Bahamian island can be identified by a distinctive kind of *Cerion*. Thus, bring me a single shell, and I can usually tell you where you spent your last vacation.

Marine species, by contrast, generally maintain much larger and more-continuous populations. Broader patterns of variation preclude any pinpoint definition of island coastlines (at Bahamian scale) by distinctive shapes or sizes of clams, snails, corals, or other oceanic forms. Terrestrial species, therefore, offer our only real hope for distinguishing islands by unique biological inhabitants. Individual Bahamian islands might house an endemic insect, or perhaps a plant, but *Cerion* surely provides the best biological marker for specifying particular locales. Insects and plants are less distinctive and harder to preserve; but if Columbus had just slipped a nearly indestructible *Cerion* shell into his vest pocket, his trusty scabbard, or his old kit bag, then we would know. Moreover, *Cerion* must have been the first terrestrial zoological object to enter Columbus's field of vision in the New World (unless a lizard darted across his path, or a mosquito drew first Caucasian blood)—though I cannot guarantee that the Admiral of the Ocean Sea had eyes to see at this scale. For *Cerion* lives right at the coastline in large populations. As they say, you can't miss 'em. And all putative landing sites on San Salvador sport large and obvious populations of my favorite snail.

While I was on San Salvador attending a biennial conference on

Bahamian Cerion *snails vary widely. A species from San Salvador's windward coast (top left) contrasts with another (top right) from the same island's leeward side—Columbus's probable landing site. Other distinctive forms are from Mayaguana (bottom left) and Turks and Caicos Islands (bottom right).*

Caribbean geology at the Bahamian Field Station, I played extensive hooky to do a survey of the local *Cerion*. San Salvador houses two major species of *Cerion*—a large, robust, whitish shell, pointed at the top, and found on promontories on the windward east coast; and a smaller, ribbier, brownish shell, barrel-shaped at the top, and found all along the leeward west coast (and most of the island interior). A single shell of either form can easily be distinguished from the characteristic *Cerion* of all other favored sites for Columbus's initial landfall.

Cerion piratarum, Mayaguana's species, belongs to the same basic group as the east-coast *Cerion* of San Salvador, but is bigger, whiter, entirely different in shape, and quite distinct from the San Salvadorian form. Similarly, *Cerion regina* of the Turks and Caicos belongs to the same general division within *Cerion,* but could not be confused with the species on San Salvador. Samana Cay, perhaps the leading alternative landing site of recent years, also houses a large and distinctive *Cerion,* easily separated from anything living on San Salvador. As a single possible exception, I confess that I could not, from one specimen, unambiguously separate the east-coast windward species of San Salvador from *Cerion fordii,* a species restricted to a few small regions of Cat Island. Columbus almost surely landed on the leeward west coast of San Salvador, however, and the *Cerion* at this site cannot be confused with the local species of any other proposed landing place. (*Cerion eximium,* the leeward, west-coast species of Cat Island, is longer, smoother, thinner-shelled, and more mottled in color than the leeward form of San Salvador.)

Erection of monuments at putative landing sites has been something of a cottage industry on San Salvador for more than a century. Three major markers now adorn the island, each located amid a large population of *Cerion* (and all shown in the preceding photographs). The *Chicago Herald* built the first monument in 1891, in preparation for the four-hundredth-anniversary celebrations of Columbus's landing and the great Chicago Columbian exposition, held a year late, in 1893. This monument, constructed largely of exotic stones and featuring a limestone globe set within the base of an obelisk, is now eroding away on the largely inaccessible promontory of Crab Cay (a two-mile walk from the nearest path, along a beautiful beach, but then up a narrow and treacherous slope). The monument sits amid one of the densest *Cerion* populations on San Salvador. I don't think that anyone would now advocate this reefy, windward site as a conceivable landing place (though the cliff might have reflected moonlight for Rodrigo's first sight of land)—yet the monument reads: "On this spot Christopher Columbus first set foot upon the soil of the New World. Erected by The Chicago Herald. June, 1891."

The other two monuments are located a mile or two apart on more plausible landing sites of the leeward west coast—both among the extensive and distinctive *Cerion* populations of this region. The second monument, placed beside an earlier obelisk erected by a yachtsman in 1951, anticipates the quincentenary of 1992 and celebrates a Japanese voyage of hope and rediscovery:

> In October 1991, a replica of the *Santa Maria*—built by the Não Santa Maria foundation of Japan—made landfall here on its journey from Barcelona, Spain, to Kobe, Japan. We came to pay homage to Columbus and his crew and to carry our message of hope for a grand harmony in the future: harmony between men and nations, between man and the environment, and between the earth and the universe.
>
> Haruo Yamamoto
> CAPTAIN, *NÃO SANTA MARIA*

The plaque on the "official" monument, a cross erected in 1954 on Long Bay, within site of the excavation that yielded the late-fifteenth-century coin, simply reads: "On or near this spot, Christopher Columbus landed on the 12th of October, 1492. Admiral Samuel Eliot Morison, USNR." The base contains yet another message of reconciliation:

> Dedication and Christmas services shared by all churches 25 December, 1956. Americans and natives worshipped together

as [a] symbol of faith, love, and unity between all nations and for
peace on earth.

And so we reach the crux of all the tension, all the triumph and tragedy,
all the drama of this great historical tale—however illuminated or alleviated
by a little side story about a distinctive land snail. We must not carp.
Columbus opened a new world, and began a process that altered human history in a permanent and fundamental way. He was a brilliant and coura-
geous sailor, and his accomplishments merit all the messages of hope and
fortitude proclaimed in unison by the monuments of San Salvador. The
messages are therefore "true" in this narrow sense—but ever so partial, and
therefore misleading as well.

As I read Columbus's log, I could thrill to his accomplishments, but I
also felt waves of revulsion at two persistent themes that never find expression on ceremonial tablets, but also set the pathways of later history. First,
his lust for gold, his almost single-minded search for the currency that
would justify his endeavor and all future exploitation. On San Salvador, he
noticed small gold rings in the noses of some Taino natives, and he persis-
tently inquired about the source. He went from island to island, looking for
mines, and thinking that he would soon encounter either the fabled golden
isle of Cipangu (Japan), or the rich courts of the grand Khan in Cathay
(China). As he visited progressively more powerful caciques (local chiefs), he
found more and more gold, but never a source area—and (obviously) never
the rich and fabled civilizations of eastern Asia. Finally, and tragically for
the local people, he did discover a source of gold on Hispaniola—and his
kinsmen built the mines that precipitated the enslavement and genocide of
the Tainos, and the total depopulation of the Bahamas.

On October 13, 1492, his second day in the New World, Columbus had
already begun his inquiries, writing in his log: "And by signs I was able to
understand that, going to the south or rounding the islands to the south,
there was a king who had large vessels of it and had very much gold." In a
classic passage, Samuel Eliot Morison writes:

A *Cerion* for Christopher

All the rest of his First Voyage was, in fact, a search for gold and Cipangu, Cathay and the Grand Khan; but gold in any event. In all else he might fail, but gold he must bring home in order to prove *la empresa* [the undertaking] a success.

In the Bahamas, his Taino guides spoke of a large nearby island called Colba (Cuba)—but Columbus heard "China" and went off in search of gold. On Cuba, he heard a rumor of gold in the island's interior at Cubanacan (meaning mid-Cuba)—but he heard *El Gran Can,* and thought that he would soon reach the imperial court. On the shore of Hispaniola, two days before Christmas, Columbus learned about gold in Cibao (the local name for central Hispaniola)—and he heard Cipangu, or Japan. But this time his countrymen would find their reward.

Second, Columbus praised the kindness and hospitality of the native Tainos. He could not have proceeded nearly so well without their enthusiastic help. Yet, his commentary speaks only about ease of domination and compulsion to service, not of gratitude or appreciation. In his very first entry for October 12, following his initial meeting and trading session with the Tainos of San Salvador, Columbus noted:

> I gave to some of them red caps and to some glass beads, which they hung on their necks, and many other things of slight value, in which they took much pleasure; they remained so much our friends that it was a marvel; and later they came swimming to the ships' boats . . . and brought us parrots and cotton thread in skeins and darts and many other things . . . everything they had, with good will.

Columbus then made an observation with practical import:

> They bear no arms, nor know thereof; for I showed them swords and they grasped them by the blade and cut themselves through ignorance; they have no iron.

And he drew a conclusion about domination, not brotherhood:

> They ought to be good servants and of good skill, for I see that they repeat very quickly all that is said to them; and I believe that they would easily be made Christians, because it seemed to me that they belonged to no religion. I, praise Our Lord, will carry off six of them at my departure to Your Highnesses, so that they may learn to speak.

Two days later, he wrote more openly about servitude: "These people are very unskilled in arms . . . With fifty men they could all be subjected and made to do all that one wished." And, from Hispaniola, near the end of the voyage, Columbus stated a plan for enslavement more explicitly: "They bear no arms, and are all unprotected and so very cowardly that a thousand would not face three; so they are fit to be ordered about and made to work, to sow and do aught else that may be needed."

And history then unfolded according to the Admiral's suggestion. The mines and estates of New Spain needed labor, and the local people, whom Columbus had called "Indians" in a mistaken belief that he had reached eastern Asia, became serfs and slaves because they could not stand against the Spanish technology of swords and gunpowder. As the natives of Hispaniola died from disease, overwork, cruelty, and (no doubt) inner distress, the Spanish governors authorized a "harvesting" of new bodies from neighboring places. And they turned to Columbus's first landfall—the Bahama islands, small bits of land with good anchorages, and unarmed people with no place to hide. In his classic book *The Early Spanish Main*, C. O. Sauer writes:

> Jamaica was known to be populous . . . Its size and tracts of difficult terrain, however, would have demanded well-organized expeditions to round up natives in number. Cuba, large and less well known, would have required even more effort. The Lucayas [Bahamas] on the other hand were a great lot of small islands, lacking refuges except by flight to another island and their people were known to be without guile; these would be the easiest to seize.

A *Cerion* for Christopher

Starting in 1509, and largely under the command of Ponce de León, the lieutenant governor of Puerto Rico, Spanish ships began to capture the Bahamian Tainos to work as slaves in Hispaniola and neighboring islands. The conquerors were thorough and rapid in their grisly work. Estimates vary, but several tens of thousands may have been thus enslaved. As the Bahamian population dwindled, the price per head rose from five to 150 gold pesos. By 1512, only twenty years after the first Columbian contact, not one Taino remained in the Bahamas. (They did not long survive in the mines of New Spain, either—and Africans were soon imported as "replacements," thus beginning another major chapter of shame in the history of the New World.) We all learned in school that Ponce de León discovered Florida in 1513 as part of a heroic and romantic quest for the Fountain of Youth. Perhaps, in part. But Ponce de León, the chief agent of Taino destruction, had sailed primarily to find a new source of slaves beyond the thoroughly depopulated Bahamian islands.

Bartolomé de Las Casas (1474–1566) began his manhood as a soldier, and sailed for Hispaniola in 1502. He participated in the conquest of Cuba and received an *encomienda* (a royal grant of land with Indian slaves). But Las Casas had a change of heart and became a priest. He preached a sermon against slavery and ill treatment of native peoples in 1514, and returned his Indian serfs to the governor. In 1515, he sailed for Spain to plead before the court for better treatment of Native Americans. He later joined the Dominican order and, in the course of a long and active life spent writing treatises and shuttling between Spain and the New World, he became a passionate and effective advocate for humane treatment of the New World's first inhabitants.

The same Las Casas copied Columbus's log to use as a source for his historical writings. As he considered Columbus's role in the story of Indian conquest and servitude, Las Casas noted the tragic beginning that might have unfolded otherwise, had only decency been able to conquer greed. Las Casas explicitly discusses the passages from Columbus's log cited earlier in this essay:

Note here, that the natural, simple and kind gentleness and humble condition of the Indians, and want of arms or protection, gave the Spaniards the insolence to hold them of little account, and to impose upon them the harshest tasks that they could, and become glutted with oppression and destruction. And sure it is that here the Admiral [Columbus] enlarged himself in speech more than he should and that what he here conceived and set forth from his lips, was the beginning of the ill usage he afterwards inflicted upon them.

As a final result, and in one of history's greatest and cruelest ironies, the first people that Europeans encountered in the New World also became the first victims of Western genocide. As one tiny consequence, no historical continuity could be maintained to preserve a human record or legend of Columbus's first landfall—and we must therefore resort to a fable about a land snail as a hypothetical way (though guaranteed for success if only Columbus had collected a single shell) to resolve this initial puzzle in the modern history of a hemisphere.

San Salvador remained uninhabited for nearly three hundred years (legends about transient pirate landings notwithstanding)—until British loyalists, fleeing the American Revolution, built plantations and imported slaves of African origin. The descendants of these slaves built the second culture of San Salvador, now vigorously in force.

We may soften the old observation that a second historical cycle often replays an initial tragedy as a derived farce. Let us only note that repeat performances tend to be more gentle. The wake of Columbus destroyed the first culture of San Salvador in the most cruelly literal way. The long arm of Columbus now threatens the second—not with death this time, but with assimilation to international corporate blandness. Most islands of the outer Bahamas remain largely "undeveloped" by modern tourism and resort culture, but the idea of Columbus's landfall provides a hook for luring people to San Salvador. After a century of small hostelries that fit well with local culture, Club Med has just built a major establishment that may change this small island into a playground of tinsel.

But many forces resist homogenization, and we should take heart. Of two examples, consider first the humor of *Homo sapiens*. A small establishment on the main road of San Salvador calls itself "Ed's First and Last Bar"—because people tend to stop by both before and after their visit to a much larger and more popular watering hole up the road a piece. But a new sign now graces the First and Last—"Club Ed," of course!

As a second example, and if only for symbolic value, consider the tenacity of *Cerion*. Club Med and its clones may one day envelop the island, sweeping up the vestiges of local culture into a modern, rootless fairyland of more gentle (and literally profitable) modern exploitation. But *Cerion* will hang tough as a marker of San Salvador's uniqueness. Unless the entire island becomes paved and manicured, *Cerion* will survive. *Cerion*, hearty and indestructable, poses no threat to agriculture or urban existence, and therefore passes largely beyond (and beneath) human notice; *Cerion* also inhabits the scrubby shoreline environments least attractive for human utility.

Cerion will survive to provide an unbroken continuity with Columbus and the original Taino inhabitants. Any snail among thousands crowded around the first Columbian monument on Crab Cay may be the great-great-great-great grandchild of a forebear that looked back at the *Pinta*—and wondered about the future in its aimless, snail-like way—when Rodrigo de Triana first raised his cry of *"Tierra!"* and altered human history forever.

1 2

THE DODO IN THE

CAUCUS RACE

Most members of my immigrant Jewish family took pride in their supposed assimilation (often more imagined than real), and derided as "greenhorns" those who stuck to old ways and tongues. Nonetheless, I well remember the lilt of Yiddish, liberally sprinkled into heavily accented English, used exclusively for a wide range of jokes and stories, or spoken as a mother tongue by the recalcitrants. In 1993, the last native Yiddish speaker of my extended family died. She was one hundred years old.

When such valued parts of natural or human diversity disappear as active, living presences, we take special interest—verging sometimes on zealous protectionism for the merest scraps—in preserving the "fossil" artifacts of extinguished vitality. And when we discover such a vestige as a pleasant and entirely accidental surprise, we feel doubly blessed for a gift bestowed by a normally uncaring world—all without our seeking, or expecting. Two recent and personal examples struck my heartstrings more than my brainstuff, and channeled my thinking to the general topic of extinction and preservation.

I saw a ten-story building, standing tall among the tenements of East Broadway on New York City's Lower East Side. I noted some Hebrew letters in raised and decaying metal along the top story. (The first *raysh* has fallen off entirely, but the outline of the Hebrew *r* can still be discerned in the incised stone beneath.) I soon recognized the word as Yiddish, not Hebrew, as I began to spell: *fay, alef, raysh* (in the incised stone) . . . *Farvarts,* or "Forward." I had found the old home of the greatest Yiddish newspaper in a once-vibrant press. Many of my relatives bought the paper daily, and I knew both the pathos and bathos of a publication that seemed almost organic in intensity—the campaigns for social justice in the sweatshops, and the *Bintel Brief* (or bundle of letters), the advice column, chockablock with questions from parents bemoaning the modern ways of their children. I felt so happy to know that the site has survived in recognizable form, even if the institution must eventually perish. (*The Forward,* now published uptown, maintains health in the altered form of weekly English and Russian editions—but the Yiddish edition drops continually in circulation, as the last speakers die.)

A few days later, I went to the movies to see *Independence Day,* the outer-space summer blockbuster of 1996. (Even the most committed intellectual can't survive on an unalloyed diet of Jane Austen remakes.) I had never noticed this unprepossessing theater on Second Avenue and Twelfth Street. But then I went inside and saw the surprising, if faded, beauty of the interior, with wonderful multicolored tilework of Moorish design. The popular film occupied the largest theater of this multiplex—the old main hall of the original building. Here, the tilework glowed in a particularly sumptuous pattern. As the film began, and the alien ships hovered over our cities, I looked up at the ceiling and noted the central pattern of dark tiles arranged in an enormous oval—almost exactly, eerily, and (obviously) quite unintentionally, mimicking the flying saucers on the screen. And then I saw the Star of David at the center of the tilework!

The Dodo in the Caucus Race

Independence Day had unfolded in the finest surviving memorial to another great institution of my ancestral culture: a Yiddish theater. I hadn't realized that any building of the old "Yiddish Rialto" on Second Avenue still existed in recognizable form, and I remembered my relatives reminiscing about the quality of a Yiddish King Lear, or the tunefulness of old Yiddish musicals. I later read that I had been visiting the Louis N. Jafee Art Theater, built in 1925, and presenting performances in Yiddish until 1945, with a brief revival between 1961 and 1965. I could only think of a favorite line from Wordsworth: "The sunshine is a glorious birth / But yet I know . . . that there hath passed away a glory from the earth."

If we regard the details of vibrant diversity as precious and glorious—and not as superfluous baubles upon Platonic essences—then the profession of preservation becomes one of the most noble callings that a person can undertake for a life's work. I shall not discuss the happy side of preservation—the restoration to vitality of lingering institutions otherwise doomed (though people who follow this calling are doubly blessed). I shall focus instead on what many people may regard as a preeminent exercise in frustration, the ultimate job for stoics, and the incarnation of numerous proverbs best represented by closing the stable door after a horse has permanently bolted: the assiduous collection, and meticulous preservation, of remains—often so few, partial, and pitiful—of people, cultures, species, and places that have permanently disappeared.

In my profession of natural history, the people charged with preserving such artifacts do their work in museums and carry the title of curator, or head (literally caretaker) of a collection. Curators do not generally enjoy high status (or salary), in large part because we unfairly devalue their activities, including the task of rescue highlighted in this essay. Our disparagement of the preservational role as either ineffably sad or almost risibly impotent (a beak in a drawer instead of ten thousand brilliantly plumed and beautifully singing birds in the bush) strikes me as greatly unfair on several grounds.

I have never met a curator who would not prefer the happier task of restoring a remnant to vitality. Nearly anyone in this line of work would take a bullet for the last pregnant dodo. But should we not admire the person who, when faced with an overwhelmingly sad reality beyond any personal blame or control, strives valiantly to rescue whatever can be salvaged, rather than retreating to the nearest corner to weep or assign fault?

Most important, the nobility of preservation arises from the nature of history itself. We needn't fret because we have no specimens of Cambrian quartz from Florida, or cannot photograph a Jurassic rainbow. Such simple objects, directly formed under unchanging laws of nature, do not vary in interesting ways from time to time or place to place. But complex objects of history, unpredictable in principle, and generated but once in all their detailed and unrepeatable glory, must utterly disappear from human understanding unless we preserve a record of their actual existence. Millions of species lived and died without leaving a single fossil sign of their residence on Earth. And we shall never greet them—a sad thought for a paleontologist with an insatiable desire to grasp the full richness of life's past. To know a physical phenomenon, we must understand the laws that govern its generation. To know a historical entity, we must preserve a record. Blessed be the recorders and collectors (see chapter 9 on a poignant and unusual case of preservation).

I want to consider the curator's role—as heroic rather than futile—in preserving the merest scraps of a record in three inaugural losses of particular symbolic importance: the extinction of the first large terrestrial mammal in 1799; the extirpation, in the 1680s, of the first animal clearly driven to death in historic times by human agency; and the genocide of the first human group encountered by Westerners in the New World, greeted in 1492 and obliterated by 1508.

Two common features intrigue me, and tell us something important about human psychology and the conceptual prejudices of Western life. First, in each case, only a paltry record could be salvaged, and all prominent preservationists focus upon this particular sadness as symbolizing the senselessness of loss. Second, and almost oddly in contradiction, all major commentators also denigrate the lost creatures as doomed by their own inadequacies—as if to expiate any guilt for the rapaciousness that made preservation necessary in the first place! Must we always blame the victims because we can't bear the truthful conclusion that baleful events really didn't need to occur? Inadequacy must lead to eventual doom, but excellence need not wither.

In 1799, a South African hunter shot the last blaauwbock, or blue antelope (*Hippotragus leucophaeus*). This species, already reduced to a tiny population living in a small area, did not come to the attention of Europeans until 1719, and did not receive a formal description until 1766. Western culture surely delivered the coup de grâce, but the blaauwbock had already been doomed either purely in the course of nature, or partly though deterioration of habitat caused by domestic sheep introduced to the region by native Africans as early as A.D. 400. This short interval of knowledge, and the animal's rarity, led to near disappearance of all palpable records. Only four mounted specimens survive in museums—the "four antelopes of the apocalypse" of my previous essay devoted to this story (reprinted in an earlier volume of essays, *Dinosaur in a Haystack*). All commentators have invoked the extreme paucity of preserved remains to carry the moral of this particular tale, and of the generality thus represented.

The first recorded extinction by human agency has become an almost automatic symbol, universally known and cited in all modes of communication—conceptually, iconographically, and even linguistically. "Dead as a doornail" only refers to immobility, for a doornail is a bolt, not a fastener. "Dead as a dodo" means totally and forever.

The Mascarene Islands—Mauritius, Réunion, and Rodrigues—located east of Madagascar in the Indian Ocean, have lost many species of birds to direct and indirect results of human activity. But the prototype and granddaddy of all extinctions also occurred here, with the death of all three species in a unique family of flightless pigeons—the solitaire of Rodrigues, last seen in the 1790s; the solitaire of Réunion (probably more closely related to the dodo), gone by 1746; and the celebrated dodo of Mauritius, last encountered in the early 1680s and almost surely extinct by 1690.

Although Portuguese sailors reached the previously uninhabited Mascarenes in the early sixteenth century, no mention of the dodo has been found before the narrative of the Dutch voyage of Jacob Cornelius Van Neck, who returned to Holland in 1599. The botanist Carolus Clusius provided the first scientific description in 1605, after observing the foot of a dodo in the home of his friend, the anatomist Peter Paauw.

Large dodos weighed more than fifty pounds. They grew a bluish gray plumage on a squarish, short-legged body, surmounted by a large head, free

From Memoir of the Dodo *(Didus ineptus), by Richard Owen, London, 1866.*

of facial feathers, and bearing a large bill with a strongly hooked tip. The wings were small and apparently useless (at least for any form of flight). Dodos laid single eggs in ground nests.

What could be easier to catch than a lumbering giant flightless pigeon? Dutch sailors didn't like the meat, and originally called the dodo a *Walgvogel,* or nauseating bird. But some portions, well cooked, tasted good enough—and no ship's victualer could afford to sneeze at such a free and bounteous supply of meat on the hoof (or vestigial wing), so to speak. Still, capture for human consumption probably didn't seal the dodo's fate, for extinction occurred primarily by indirect effects of human disturbance. Early sailors brought pigs and monkeys to the Mascarenes, and both multiplied prodigiously. Both species apparently feasted on dodo eggs, easily acquired from the unprotected ground nests—and most naturalists attribute a greater proportion of deaths to these imports than to direct human action. In any case, no one ever saw a live dodo on Mauritius after the early 1680s. In 1693, the French explorer Leguat spent several months on Mauritius, looked hard for dodos, and found none.

Dodos provide a particularly good illustration for my two conflicting principles: lament at the paucity of preserved remains, and blame for death largely laid to the victim's inadequacy. Human contact may have lasted less than a century, but dodos were both locally abundant and well documented. In this context, remarkably little remains in our museums as a testimony to this prototype of all extinctions. Several seventeenth-century paintings and drawings exist, some made in Europe, and apparently from life. We have no absolute proof that living dodos ever arrived in Western nations, but strong circumstantial evidence suggests that nine or ten birds might have come to Holland, two to England, one to Genoa, two apparently to India, and one, perhaps, even to Japan. H. E. Strickland, author of the classic 1848 monograph on the dodo, spoke of this paucity of evidence:

> We possess only the rude descriptions of unscientific voyagers, three or four oil paintings, and a few scattered osseous fragments, which

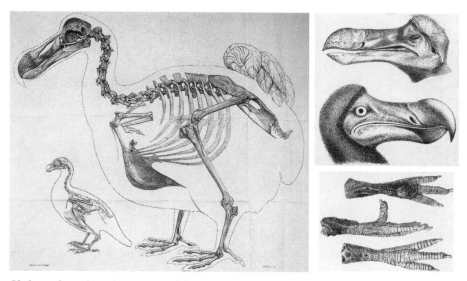

Skeleton from Owen's Memoir of the Dodo. *Feet and skull from* The Dodo and Its Kindred, *by H. E. Strickland and A. G. Melville, 1848.*

have survived the neglect of two hundred years. The paleontologist has, in many cases, far better data for determining the zoological characters of a species which perished myriads of years ago.

A few partial skeletons and many scattered bones, most excavated from Mauritian bogs after 1850, now grace our museums, but remarkably little evidence remains from birds that humans saw alive. Copenhagen has a skull, Prague a bit of a beak. Of flesh and blood, we have only one preserved foot in the British Museum, and a head and foot in Oxford. What a paltry legacy for an animal that occupies such a central place in our legends and history!

The tale of the last dodo is especially poignant. A complete stuffed specimen existed in the collection of John Tradescant, developer of the first important English museum of natural history. Tradescant bequeathed his collection to Elias Ashmole, who then founded the Ashmolean Museum at Oxford University. There the specimen languished and rotted away until, in 1755, the directors of the museum consigned the "LAST OF THE

The Dodo in the Caucus Race

DODOS" to the flames (to cite Strickland's words in his own upper case). An astute curator managed to save the head and one foot—virtually the only fleshly evidence now existing for the first animal driven to extinction by modern humans. Nearly a century later, the great geologist Charles Lyell described this desecration in words redolent of pain, and expressing the solemn duty of all true curators: to preserve the remains when we cannot rescue the living, and to maintain the records when we cannot even conserve the remains—lest we forget, lest we forget:

> Some have complained that inscriptions on tombstones convey no general information except that individuals were born and died—accidents which happen alike to all men. But the death of a species is so remarkable an event in natural history, that it deserves commemoration; and it is with no small interest that we learn from the archives of the University of Oxford the exact day and year when the remains of the last specimen of the Dodo, which had been permitted to rot in the Ashmolean Museum, were cast away.

Strickland used the same argument to justify the time and expense of publishing a monograph on "the first clearly attested instance of the extinction of organic species through human agency." We may well mark his prophetic words in our current age of greatly accelerated anthropogenic extinction:

> We cannot see without regret the extinction of the last individual of any race of organic beings, whose progenitors colonized the pre-Adamitic Earth . . . The progress of Man in civilization, no less than his numerical increase, continually extends the geographical domain of Art by trenching on the territories of Nature, and hence the zoologist or botanist of future ages will have a much narrower field for his researches than that which we enjoy at present. It is, therefore, the duty of the naturalist to preserve to the Stores of Science the knowledge of these ancient and expiring organisms, when he is

unable to preserve their lives; so that our acquaintance with the marvels of Animal and Vegetable existence may suffer no detriment by the losses which the organic creation seems destined to sustain.

Yet, for all these expressions of sadness and determination, few naturalists ever spoke well of the poor dodo while it lived, or even later when theory demanded a rationale for the dodo's extinction, and "blaming the victim" seemed an easier course than admitting an eminently avoidable tragedy. What creature has ever been subjected to more ridicule and derision? To be sure, the dodo was not a lovely creature by our conventional standards of beauty. The bird seemed inept, again by our inappropriate criteria—a waddling creature, incapable of flight and condemned to raising nestlings on the open ground. But have we not been taught to look behind overt appearance? Could we not, in words of the great British anatomist Richard Owen, champion "the beauty of its ugliness"?

On the contrary, we did nothing but deride and stigmatize. Diverse theories for the dodo's etymology agree on only one point: whatever the derivation, the intent was surely pejorative. Some ascribe *dodo* to a Portuguese word for "foolish" (unlikely, since the few Portuguese sailors to the Mascarenes never mentioned dodos). Others derive the name from *dodoor*, a Dutch word for "sluggard." Most seventeenth-century sources cite some orthographic variant of *dodaers*—the name generally used by Dutch sailors, and meaning, roughly, "fat ass." Moreover, the official scientific names show no more *gentilesse*. Linnaeus called the species *Didus ineptus*—*Didus* as a Latinization of *dodo,* and *ineptus* for obvious reasons. Modern ornithologists often use the earlier name *Raphus,* given by the naturalist Moehring as a Latinization of the Dutch *reet,* a vulgar term for "rump."

From the very beginning, even while the dodo lived in prosperity on Mauritius, European descriptions dripped with disdain. For example, in 1658, the naturalist Bontius began the tradition of blaming the victim, even before the eventual outcome, by linking the dodo's deficiencies to ease of capture: "It hath a great, ill-favored head . . . It is a slow-placed and stupid bird, and which easily becomes a prey to the fowlers."

After 1690, the chorus of disdain only increased, for now the dodo could be blamed for its singular fate. Consider the mid-eighteenth-century description of that ultimate arbiter of taste in science, the preeminent naturalist Georges Buffon, best remembered today in general culture for his motto, *"le style c'est l'homme même"* (the style is the man himself). Buffon, as cited in chapter 20, regarded the sloth as a prototype of ugliness and inadequacy among mammals. Thus, in labeling the dodo as a sloth among birds, Buffon could not have made his judgment more clear or cutting:

> The body is massive and almost cubical; it is scarcely held up by two fat and short legs. The head is so extraordinary that one might take it for a fantasy of a painter of grotesques. This head, mounted on a thick and goiterous neck, consists almost entirely of an enormous beak . . . All this results in a stupid and voracious appearance . . . Its heaviness, which usually presupposes strength in animals, here only produces lethargy . . . The dodo is, among birds, what the sloth represents among mammals: one might say that this bird is made of brute and inactive matter, where the vital molecules are too sparse. It has wings, but the wings are too short and too weak to raise it into the air. It has a tail, but the tail is disproportionate and out of place. One might take it for a tortoise decked out in the covering of a bird—and nature, in giving it such useless ornaments, almost shows a desire to add embarrassment to bulk, clumsiness [*gaucherie* in the original French, almost an English word now as well, and referring literally to left-handedness] of motion to the inertia of mass, and to give the creature a gross heaviness all the more shocking when we realize that it is a bird.

Interestingly, only H. E. Strickland, the dodo's most assiduous student and monographer, spoke well of the bird in his 1848 treatise. We may disdain this creature by our own standards, for even Strickland admitted that "we must figure it to ourselves as a massive clumsy bird, ungraceful in its form, and with a slow waddling motion." But who are we to judge, when God created each animal with optimal features for its own designated mode of life:

> Let us beware of attributing anything like imperfection to these anomalous organisms, however deficient they may be in those complicated structures which we so much admire in other creatures. Each animal and plant has received its peculiar organization for the purpose, not of exciting the admiration of other beings, but of sustaining its own existence. Its perfection, therefore, consists, not in the number or complication of its organs, but in the adaptation of its whole structure to the external circumstances in which it is destined to live. And in this point of view we shall find that every department of the organic creation is equally perfect.

But, even more interestingly, Strickland still felt that he had to find a rationale based on inevitability, rather than contingent and preventable despoliation, for the dodo's extinction. So he argued that species, like individuals, probably go through a determined cycle of birth, maturation, and death—and that humans therefore only hastened an ineluctable end:

> It appears, indeed, highly probable that Death is a law of Nature in the Species as well as in the Individual; but this internal tendency to extinction is in both cases liable to be anticipated by violent or accidental causes. Numerous external agents have affected the distribution of organic life at various periods, and one of these has operated exclusively during the existing epoch, *viz.* the agency of Man.

But Richard Owen, England's finest anatomist, would not let Strickland get away with such mush. In his own 1866 monograph on the dodo, Owen reasserted the bird's inherent inferiority in absolute terms. Citing the justice of Linnaeus's name, Owen wrote:

> The brain is singularly small in the present species of *Didus;* and if
> it be viewed as an index of intelligence of the bird, the latter may
> well be termed *ineptus.*

Owen then attributed the dodo's degeneration to an easy life on Mauritius,
an island free of predators and competitors:

> That there would be nothing in the contemporaneous condition of
> the Mauritian fauna to alarm or in any way to put the Dodo to its
> wits; being, like other pigeons, monogamous, the excitement, even,
> of a seasonal or prenuptial combat, might, as in them, be wanting.
> We may well suppose the bird to go on feeding and breeding in a
> lazy, stupid fashion, without call or stimulus to any growth of cere-
> brum proportionate to the gradually accruing increment of the bulk
> of the body.

Owen then specifically attacked Strickland's notion of universal appro-
priateness and local perfection, citing the theories of two great French naturalists, Buffon and Lamarck, to support his notion of genuine degeneration:

> The Dodo exemplifies Buffon's idea of the origin of species through
> departure from a more perfect original type by degeneration; and
> the known consequences of the disuse of one locomotive organ and
> extra use of another indicate the nature of the secondary causes that
> may have operated in the creation of this species of bird, agreeably
> with Lamarck's philosophical conception.

Finally, Owen fired his ultimate salvo: Does not the simple fact of
extinction, all by itself and *tout court,* seal the case for inadequacy?

> Nevertheless the truth, as we have or feel it, should be told. In the
> end it may prove to be the more acceptable service. The *Didus inep-
> tus,* through its degenerate or imperfect structure, howsoever
> acquired, has perished.

So, too, did the first human group encountered by Europeans in the New World—also on islands—perish quickly by rapacity, exploitation, and the sword. The previous essay tells a sad tale of the Bahamian Tainos, met by Columbus on October 12, 1492—and fully extirpated, following forced removal and indenture in Hispaniola, by 1508. Columbus spoke well of the physical appearance of the Bahamian Tainos, admiring their large stature and attractive appearance—"their forms being very well proportioned, their bodies graceful, and their features handsome," as he wrote in his log. Yet Columbus also noted the ease of their potential exploitation: "They do not carry arms and have no knowledge of them. They have no iron . . . With fifty men they could all be subjected and made to do all that one wished." Columbus collected nothing on the Bahamas—not even the single *Cerion* shell that could have resolved the issue of his landfall (see chapter 11)—so posterity received no legacy from the native Bahamians beyond a verbal record.

During the 1880s, the Western world, at the height of colonial expansion, and still untroubled by a history of exploitation (even genocide) against "inferior" peoples of other cultures, began to gear up for celebrations to mark the four hundredth anniversary of Columbus's landfall. At the same time, one of my favorite scientists, Louis Agassiz's last student, visited the Bahamas to pursue his work on the anatomy and embryology of marine invertebrates. As a man of general curiosity, W. K. Brooks, professor of zoology at Johns Hopkins University, turned his attention to other aspects of local natural history. He contemplated the fate of the original inhabitants, and he discovered that no anatomical remains had ever been recorded. He began to make inquiries and found that a few skeletons had been recovered from caves, but never properly described. Brooks secured the cooperation of local collectors, and studied this paltry legacy of the vibrant and complex culture first met by Europeans. Brooks published his results, the only anthropological research he ever pursued, in a technical article for the *Memoirs of the National Academy of Sciences* (1889), and in a general article for *Popular Science Monthly* in the same year.

Brooks began his popular article by making the link to forthcoming

Columbian celebrations, and with passionate lament for a destruction so brutal, and so total, that only one legacy of the original Bahamian culture survives—as a disembodied word, not even a palpable thing!

> In three years the world will unite in celebrating the four hundredth anniversary of what from our point of view, is the grandest and most important event in history, the landing of Columbus; but in our consciousness of its profound significance, are we not in danger of forgetting that the Spaniards discovered America in the way that pirates discover a vessel with a helpless crew? . . . [They] found the Bahamas in the possession of a prosperous and happy people . . . Twelve years afterward every soul of the population of more than forty thousand men, women, and children had perished in a strange land under the lashes of the slave-driver; the race was blotted off the face of the earth, and the only impression which has been left upon our civilization by those who first welcomed it to this continent is a single word, which, together with the luxurious article it designates, has spread over the whole earth. [They] gave us the hammock, and this one Lucayan word is their only monument.

(A few other words, including *tobacco,* derive from the same language group. But Columbus first encountered tobacco on Hispaniola, and *hammock* entered Western languages as a unique Bahamian contribution.)

Following the general pattern featured as the theme of this essay, Brooks then located the focus of tragedy in the extreme paucity of remains, and expressed special pleasure in the task of rescue:

> All traces of their existence were almost completely obliterated by the conquerors . . . The Spaniards had not time nor inclination for the study of anthropology, and their random notes give us little or no knowledge of the people they destroyed, and I was therefore greatly pleased when I obtained in the Bahamas . . . the material for a satisfactory study of their anatomical characteristics.

But then, and also following the standard pattern, Brooks larded his dry anatomical descriptions with statements of disparagement—as if to suggest that the native Bahamians had been doomed by their own inherent inferiority. He found, or so he thought, two signs of biological lowliness. First, he stated a claim for similarity between primitive races and lower mammals: "Certain variations in human crania, which are exceptions in man, but normal in certain other mammals, occur more commonly in savage than in civilized races." Then, despite the smallness of his sample, Brooks claimed confirmation for this principle:

> The four Lucayan skulls, however, present two cases or 50 per cent of triquetral bones in the lambdoidal suture, and as there is no reason for attaching any particular morphological importance to this peculiarity, it seems probable that savages or primitive races may be more variable or irregular as regards their osteological characteristics than civilized races.

Second, Brooks interpreted several features of the skull as "bestial," even while generally affirming Columbus's impression of good stature and fair form (see chapter 11). He wrote with more dispassion in his technical work: "The muscular attachments on the occipital and those on the mandible, and the great overhanging superciliary [brow] ridges give to these skulls a bestial expression and indicate that their possessors must have been unusually muscular men." But he stated with more fervor and prejudice in his popular article:

> They had protuberant jaws and the powerful neck and jaw muscles of true savages, and the outlines of the skulls have none of the softness and delicacy which characterizes those of more civilized and gentle races of men.

I confess that I do have trouble in reconciling these two invariant and contradictory themes of early literature on preserving the remains of our initial depredations: the fervor and nobility of rescue, even for the merest scraps;

with disparagement of the creatures thus preserved as artifacts, and an attribution of their extinction, in large part, to these supposed inadequacies—for why should we struggle to preserve the inept with such zeal? Still I do not doubt—and I certainly do honor—the genuine feelings of scientific achievement and moral fulfillment that attended the rescue of paltry artifacts as unique remembrances. W. K. Brooks expressed the psychological dimension particularly well when he wrote of the inspiration provided by genuine objects, rather than replicas or mere words:

> There is not much intrinsic interest in a few fragments of human bones, but the Lucayan [native Bahamian] skull which stands upon my table as I write gives life and vivid reality to the familiar story ... and calls up in all its details with startling clearness the drama of the Bahama Islands.

As for the tendency to disparage, I suggest that we need new concepts and metaphors to replace the false and constraining notions, however comforting, of predictable progress in the history of life (with sad, but inevitable loss of inferior creatures), and sensible causality for all major events. Fortunately, we may find a wonderful example and opportunity for correction in the most famous literary appearance of the dodo.

Lewis Carroll viewed himself as a bumbling and ungainly man, and therefore strongly identified with the dodo. In chapter three of *Alice in Wonderland,* after all characters have become thoroughly soaked, a long and vociferous argument breaks out about the best mode of getting dry. Finally, the dodo suggests a resolution. "I move," he states, "that the meeting adjourn, for the immediate adoption of more energetic remedies." "The best thing to get us dry," the dodo continues, "would be a Caucus-race." The dodo therefore lays out a circular course, and places all the participants at random starting points:

> There was no "One, two, three, and away!", but they began running when they liked, and left off when they liked, so that it was not easy

Drawing by John Tenniel from Alice's Adventures in Wonderland, *by Lewis Carroll, 1865.*

to know when the race was over. However, when they had been running half an hour or so, and were quite dry again, the Dodo suddenly called out "The race is over!"

The participants remain puzzled and ask, "But who has won?"

This question the Dodo could not answer without a great deal of thought, and it stood for a long time with one finger pressed upon its forehead (the position in which you usually see Shakespeare, in the pictures of him), while the rest waited in silence. At last the Dodo said "Everybody has won, and all must have prizes."

I suspect that life runs more like a caucus race than along a linear course with inevitable victory to the brave, strong, and smart. If we truly embraced this metaphor in conceptual terms, we might even be able to adopt a better position for considering the moral consequences of human actions, as suggested by Lewis Carroll's wise dodo: no judgments of superiority or inferiority among participants; no winners or losers; and cooperation with ends attained and prizes for all. (No one wants a caucus race for all human activities, of course. Some people do play the piano, or hit home runs, better than others—and such achievements deserve acknowledgment and reward. But when we talk about the intrinsic and ultimate worth of a human life, the judge of the caucus race becomes the wisest of men.)

And finally, speaking of races, let us not forget the most famous statement in our literature about the salutary humility—even the resulting freedom—we might obtain by admitting that the universe does not respect our preferences, and often operates on random pathways with regard to our hopes and intentions. The death of the dodo really doesn't make sense in moral terms, and didn't have to occur. If we own this contingency of actual events, we might even learn to prevent the recurrence of undesired results. For the Preacher of Ecclesiastes wrote: "I returned, and saw under the sun, that the race is not to the swift, nor the battle to the strong . . . but time and chance happeneth to them all."

13

THE DIET OF WORMS AND THE DEFENESTRATION OF PRAGUE

I ONCE ATE AN ANT (CHOCOLATE COVERED) ON A DARE. I HAVE NO AWFUL memories of the experience, but I harbor no burning desire for a repeat performance. I therefore feel poor Martin Luther's pain when, at the crux of his career, in April 1521, he devoted ten days to the Diet of Worms (washed down with a good deal of wine, or so I read).

I am a collector by nature, and mental drawers have more room for phrases and facts than physical cabinets maintain for specimens. I therefore reserve one cranial shelf for the best funny or euphonious phrases of history. "The Diet of Worms" remains my prize specimen, but I award second place to another *D*-phrase of European history: "the Defenestration of Prague" in 1618—the "official" trigger of the Thirty Years War, one of the most extended, horrendous, and senseless conflicts in Western culture.

I do not believe in vicarious experience and will go to great, even absurd, lengths to stand on the true spot, or place a hand on the very wall. I could have written *Wonderful Life* without a visit to the Burgess Shale, but what a sacrilege! Walcott's fossil quarry is holy ground, and only a four-mile trek from the main road.

I therefore accepted a recent invitation for a lecture in Heidelberg on the stipulation that my hosts drive me to nearby Worms, site of the Diet. (I had, three years earlier, stood on the square in Prague where those bodies once landed after ejection from an upper-story window.) Now, with pilgrimages completed to the sources of both phrases that most caught my fancy in Mrs. Ponti's fifth-grade European history class, I can muse more formally upon the sadly common theme behind the two *D*'s—our cursed tribal tendency to factionalize, fight, and then, so often in our righteous certainty, to define our opponents as vermin and try to expunge either their doctrines (by censorship and fire) or their very being (by genocide). The Diet of Worms and the Defenestration of Prague mark two cardinal events in the sad chronology of hatred and bloodshed surrounding a central theme of Western history, one filled with aspects of grandeur as well—the schism of "universal" Christianity into Catholic and Protestant portions.

The Diet, or governing body, of the Holy Roman Empire met at the great medieval Rhineland city of Worms in 1521, partly to demand the recantation of Martin Luther. (German sources call the Diet a *Reichstag*. Moreover, German fishbait is spelled with a "u," not an "o" as in English. Thus, the *Reichstag zu Worms* packs no culinary punch in the original vernacular.)

In school, I learned the heroic version of Luther before the Diet of Worms. This account (so far as I know, and have just affirmed by reading several recent biographies) reports factual material in an accurate manner—and is therefore "true" in one crucial sense, yet frightfully partial and therefore misleadingly incomplete in other equally important ways. Luther, excommunicated by Pope Leo X in January 1521, arrived in Worms under an imperial guarantee of safe conduct to justify or recant his apostasies before the militantly Catholic and newly elected Holy Roman Emperor, Charles V, heir to the Hapsburg dynasty of central Europe and

Martin Luther defending himself before Holy Roman Emperor Charles V at the Diet of Worms, April 1521.

Spain, and twenty-one-year-old grandson of Ferdinand and Isabella, monarchs of Spain, and patrons of Christopher Columbus.

Luther, with substantial support from local people of all classes, including his most powerful protector, Frederick the Wise, Elector of Saxony, appeared before Charles and the Imperial Diet on April 17. Asked if he would retract the contents of his books, Luther begged some time for consideration (and, no doubt, for preparation of a rip-roaring speech). The emperor granted a one-day recess, and Luther returned on April 18 to make his most famous statement.

Speaking first in German and then in Latin, Luther argued that he could not disavow his work unless he could be proved wrong either by the Scriptures or by logic. He may or may not have ended his speech (reports vary) with one of the most famous statements in Western history: *Hier stehe ich; ich kann nicht anders; Gott helfe mir; Amen*—Here I stand; I cannot do otherwise; God help me; Amen.

Faced with Luther's intransigence, the Emperor and a rump session of the Diet issued the Edict of Worms on May 8. But that document, banning Luther's work and enjoining his detention, could not be enforced, given the strength of Luther's local support. Instead, under Frederick's protection, Luther "escaped" to the castle of Wartburg, where he translated the New Testament into German.

A stirring story, invoking some of the finest themes in Western liberal and intellectual traditions: freedom of thought, personal bravery against authority, the power of one man with a grand idea before the crumbling weight of centuries. But dig just a little deeper, below the overt level of hagiography and school-day moralisms, and you enter a quagmire of intolerance and mayhem on all sides. Scratch the surface of soaring notions like "justification by faith," and you encounter a world where any major idea becomes a political instrument in a quest for social order, or a tool in the struggle for power between distant popes and local princes. Consider the operative paragraph of the Edict of Worms, complete with a closing metaphor about diets in the modern culinary sense:

> We want all of Luther's books to be universally prohibited and forbidden, and we also want them to be burned . . . We follow the very praiseworthy ordinance and custom of the good Christians of old who had the books of heretics like the Arians, Priscillians, Nestorians, Eutychians, and others burned and annihilated, even everything that was contained in these books, whether good or bad. This is well done, since if we are not allowed to eat meat containing just one drop of poison because of the danger of bodily infection, then we surely should leave out every doctrine (even if it is good) which has in it the poison of heresy and error, which infects and corrupts and destroys under the cover of charity everything that is good.

These words may be chilling enough when confined to the destruction of documents. But annihilation often extended to the inventors of unorthodoxies, and to the genocide of followers. Of the early heretics mentioned above, Priscillian, bishop of Avila in Spain, was convicted of sorcery and immorality, and executed by the Roman emperor Maximus in 385. The later Albigensians fared far worse. These ascetic communitarians of southern France frightened papal and other authorities with their views on the corruption of clergy and secular rulers. In 1209, Pope Innocent III urged a crusade against them—just one among so many examples of Christians annihilating other Christians—and the resulting war effectively destroyed the Provençal civilization of southern France. The Inquisition mopped up during the next several decades, thus completing the extirpation of an unpopular view by genocide. Grisly, but effective. The *Encyclopaedia Britannica* simply states: "It is exceedingly difficult to form any very precise idea of the Albigensian doctrines because present knowledge of them is derived from their opponents."

If Luther and other reformers had promoted their new versions of Christianity in the name of love, toleration, and respect, then I might accept the heroic version of history as progress inspired by rare individuals of broader vision. But Luther could be just as dogmatic, just as unforgiving, and just as bloodthirsty as his opponents—and when his folks took the reins of power, the old tactics of banning, book burning, and doctrinal murder continued. For example, Luther had originally held little animus toward Jews, for he hoped that his reforms, by eliminating papal abuses, might lead to their conversion. But when his hopes withered, Luther turned on his vitriol and, in a 1543 pamphlet titled *On the Jews and Their Lies,* recommended either forced deportation to Palestine, or the burning of all synagogues and Jewish books (including the Bible), and the restriction of Jews to agrarian pursuits.

In his most horrific recommendation (and on the eve of supposed personal happiness in his marriage to Katherine von Bora), Luther advocated the wholesale slaughter of German peasants, whose rebellion had recently been so brutally suppressed. Luther had his reasons and frustrations, to be sure. He had never supported uprising against secular authority, although some of the more moderate peasant groups had used his teachings as justifications. Moreover, the militant faction of peasants had been led by his bitter theological enemy, Thomas Müntzer. Political conservatives like Luther always take a dim view (if only to save their own skins) of insurrections by large and poorly disciplined groups of disenfranchised people, but Luther's recommendations for virtual genocide, as presented in his tract of 1525 *Against the Murderous and Thieving Hordes of Peasants,* makes my skin crawl, especially as a recommendation (however secular) from a supposed man of God:

> If the peasant is in open rebellion, then he is outside the law of God . . . Rebellion brings with it a land full of murders and bloodshed, makes widows and orphans, and turns everything upside down like a great disaster. *Therefore, let everyone who can, smite, slay, and stab, secretly or openly,* remembering that nothing can be more poisonous, hurtful, or devilish than a rebel. It is just as when one must kill a mad dog; if you don't strike him, he will strike you, and the whole land with you [my italics].

The victorious nobility followed Luther's recommendations, and estimates of the death toll (mostly inflicted upon rebels who had already surrendered and therefore posed no immediate threat) range to 100,000 people.

Sad tales of mass murder perpetrated by differing factions of a supposedly united cause haunt human history. I don't think that Christians are worse than other folks in this regard; we just know these stories better as defining incidents of a culture shared by most readers of this book. I am not speaking of isolated executions, but of wholesale slaughters, however unknown to us today for two eerie reasons. First, Hitlers of the past didn't possess the technology (though they probably had the will) to kill six million in a few years, so their depredations, though thorough, were more local. Second, obliterated cultures of bygone times featured fewer people, living in limited areas, and publishing little or no documentation. An older style of genocide could therefore be devastatingly complete and effective, truly wiping out all memory of a vibrant people.

I have already mentioned the Albigensian crusade. In 1204, the Fourth Crusade, having failed to reach Palestine through Egypt to conquer the Holy Land, sacked the Byzantine Christian capital of Constantinople instead, imposing more mayhem upon people and art than the "infidel" Ottomans exacted when the city finally fell from Christian rule in 1453. The rupture of Europe into Protestant and Catholic parts provided more opportunity for such divisive destruction—and Luther's legacy surely includes as much darkness as light. Which brings me to the Thirty Years War and the Defenestration of Prague.

Throwing people out of windows has a long legacy in this beautiful city—a true Scandal in Bohemia, so to speak. In each major incident but the last, rebelling Protestants (or proto-Protestants) tossed entrenched Catholics out of their strongholds. The "official" Defenestration (with a capital *D*) occurred in 1618. Local Protestants, justifiably enraged when the very Catholic King Ferdinand II reneged on promises of religious freedom, stormed Hradčany Castle and threw three Catholic councilors out of the window and into the moat. (Legend states that they walked away, embarrassed but unharmed, thanks either to good fortune or to the good aim of their adversaries—for they landed in a large and soft dunghill.)

The rebels of 1618 had consciously reenacted a past incident that they wished to claim as part of a proud and continuous history. (A Latin window, by the way, is a *fenestra*—so *defenestration* is just a fancy word for throwing something out such an opening.) The memory of Bohemian religious reformer Jan Hus, burned for heresy in 1415 and claimed by later Protestants as a precursor, inspired the initial defenestration of Prague in 1419. A Hussite army (if you like them) or a rabble (if you don't) stormed the New Town Hall and threw three Catholic consuls and seven citizens out the window (some to their death, for no cushioning dunghill broke these falls), and Bohemia did pass to Hussite rule for a time. Yet another, but lesser known, defenestration occurred in 1483. King Vladislav had restored Catholic dominion, so another dissident band of Hussites threw the Catholic mayor out the window.

The tragic epilogue to this sequence will be remembered by readers just a few years older than me. Jan Masaryk, son of Thomas Masaryk, the founder of the Czech Republic, continued to serve as the only noncommunist minister of the postwar puppet government. On March 10, 1948, his body was found in the courtyard of Czernin Palace. He had fallen to his death from a window forty-five feet above. Had he jumped as a suicide (with ironic consciousness of his nation's history), or had he been pushed in a murder? The case has never been solved.

The Protestant triumph after the official defenestration lasted only two years and ended in yet another splurge of murder and destruction. With powerful Hapsburg support, Catholics regrouped, and decisively defeated the Protestants in the Battle of the White Mountain on November 8, 1620. Weeks of plunder and pillage followed in Prague. A few months later, twenty-seven nobles and other citizens were tortured and executed in the Old Town Square. The victors hung twelve heads, impaled on iron hooks, from the Bridge Tower as a warning.

The resulting Thirty Years War cannot be reduced to a dichotomous struggle between Catholics and Protestants—but this essential division did define much of the temper and zealotry of the controversy (for we seem able to kill "apostates" far more easily than merely errant compatriots). Much of central Europe lay in ruins, as the mercenary armies of various potentates ravaged their way through the countryside, burning, raping, and pillaging as they went. Nor did the battles of Protestants against Catholics end with the Treaty of Westphalia in 1648. The ruined castle of Heidelberg, beautifully lit at night, turns the entire city into a romantic stage set for *The Student Prince*. But Heidelberg retains no medieval buildings, and the castle lies in ruins, as the consequence of yet another disastrous internecine war among Christians—when the Protestant Elector of the Rhineland Palatinate (with Heidelberg as a center) died without heir, and Catholic

France claimed the territory because the Elector's sister had married Philip of Orléans, brother of Louis XIV.

A somewhat cynical, but sadly accurate, principle of human history states that when things look bad, they can still get far worse. If Christians could slaughter each other with such gusto and ferocity, what could true outsiders expect—for non-Christians could be defined even more easily as beyond human worth and therefore ripe for elimination. For this final chapter in man's inhumanity, we turn to the obvious test: the fate of Jewish communities in medieval and Renaissance Europe at the time of diets and defenestrations.

Jewish settlements persisted for one thousand years in the Rhineland. Every city that I visited—Worms, Speyer, Rothenburg—maintains memorials to the persecution and elimination of these communities, while tourist outlets sell long and informative pamphlets on their history, lest we forget. One almost feels a concerted and commendable attempt to expiate what cannot be undone—a tale of prolonged intolerance capped by such recent memory of the brutally and entirely effective last curtain (at least locally) of the "final solution."

Gershom ben Judah, known as the "light of the exile," headed the rabbinic academy of Mainz at the end of the tenth century, before the first millennium of our era's common time had passed. His most celebrated disciple, the great Talmudist Rashi, studied in Worms around 1060. Rashi's most noted follower, Meir Ben Baruch (known to pious Jews by his acronym as the Maharam), headed the Jewish community in Rothenburg, now the most perfectly preserved, entirely walled, and touristically flooded of all medieval towns. In 1286, the Emperor Rudolph I abrogated the political freedom of Jews and imposed special taxes to make these despised people *servi camerae* (or serfs of the treasury). Rabbi Meir tried to lead a group of Jews to Palestine, but he was arrested and confined in an Alsatian fortress. His people raised an enormous ransom, but Meir refused (and died in prison) because he knew that a purchased freedom would only encourage the emperor to capture other rabbis for revenue. Fourteen years later, a Jewish merchant in Worms ransomed the great rabbi's body. His tomb, and that of the merchant, occupy adjacent places in the Jewish cemetery of Worms. Following an ancient custom, Jewish visitors and residents (mostly Russian émigrés) still write their prayers and requests on scraps of paper and place them, weighted down by small stones, atop the Maharam's tomb.

After Meir's exile, the Jews of Rothenburg were expelled to a ghetto beyond the city walls, and then, in 1520, banished entirely and forevermore. Only the small dance hall remains (because it became a poorhouse for Christians), with a few tombstones in Hebrew, mounted on the garden wall.

The larger Jewish community of Luther's Worms survived longer, but just as precariously. In 1096, soldiers of the First Crusade passed through Worms and ravaged the Jewish quarter. In 1349, nearly all the Jews of Worms were murdered on the false accusation that they had brought the plague by poisoning the wells. In 1938, on the infamous *Kristallnacht,* the Jewish synagogue burned to the ground. More than one thousand Jews of Worms perished in the Holocaust. The reconstructed synagogue now serves as the centerpiece of a Jewish museum and memorial, but not as a place of worship, for no active Jewish community now exists in Worms.

Two plaques on the synagogue wall tell a tale of hope and despair. The first, mounted shortly after the end of World War II, contains the names of Jewish citizens presumed dead in the Holocaust. Happily, some of these people had survived (in refugee camps, unknown to makers of the plaque). Their raised bronze names have been filed off, leaving blank spaces of victory. But further records of the Holocaust then documented more deaths, and these names adorn the second plaque—greatly exceeding in number the names happily erased from the first memorial.

Ironically, only the Jewish cemetery survived intact, thanks to a ruse (according to local tradition) of the town archivist, a sincere Christian with great respect for Jewish traditions. Himmler had expressed a passing interest in the cemetery during a prewar visit. When local Nazis later ordered the destruction of the cemetery (located on the other side of town, beyond the walls), the archivist exaggerated Himmler's casual comment into an explicit order for preservation. Cautious local authorities never checked with Berlin—and a place of death remains as the only unscathed survivor of a millennium's existence for one of Europe's most illustrious Jewish communities.

If you have been wondering why I recount these tales from the dark side of human history in an essay on evolutionary biology, I do intend to segue toward an ending on both a positive and a Darwinian note. Humans are capable of such glory—and such horror: the pogroms of Worms, and Luther's stirring speech at the Diet of Worms; the numerous defenestrations of Prague, and the magnificent baroque architecture of Prague. We bask in the glory with simple pleasure; but we contemplate the horror with anguish and puzzlement—and with a burning urge to explain how creatures capable of such decency can promote such iniquity of their own free will (and with apparent moral calm and intensity of supposed purpose).

But do we perpetrate the darkness "of our own free will"? Perhaps the most popular of all explanations for our genocidal capacity cites evolutionary biology as an unfortunate source—and as an ultimate escape from full moral responsibility. Perhaps we evolved these capacities as active adaptations now gone awry in the modern world. Current genocide may be a sad legacy of behaviors that originated for Darwinian benefit during our ancestral construction as small bands of hunters and gatherers on the savannahs of Africa. Darwin's mechanism, after all, encourages only the reproductive success of individuals, not the moral dream of human fellowship across an entire species. Perhaps the traits that lead to modern genocide—xenophobia, tribalism, anathematization of outsiders as subhuman and therefore subject to annihilation—rose to prominence during our early evolution because they enhanced survival in tiny, nontechnological societies based on kinship and living in a world of limited resources under a law of kill-or-be-killed.

A group devoid of xenophobia and unschooled in murder might invariably succumb to others replete with genes to encode a propensity for such categorization and destruction. Chimpanzees, our closest relatives, will band together and systematically kill the members of adjacent groups. Perhaps we are programmed to act in such a manner as well. These grisly propensities once promoted the survival of groups armed with nothing more destructive than teeth and stones. In a world of nuclear bombs, such unchanged (and perhaps unchangeable) inheritances may now spell our undoing (or at least propagate our tragedies)—but we cannot be blamed for these moral failings. Our accursed genes have made us creatures of the night.

This superficially attractive balm to our collective conscience is nothing but a cop-out based on deep fallacies of reasoning. (Perhaps the tendency to think by such fallacies represents our real evolutionary legacy—but this is another speculation for another time.) I am quite happy to acknowledge that we have a biologically based capacity to categorize humans as insiders or outsiders, and then to view outsiders as beyond fellowship and ripe for slaughter. But where can such an argument take us in terms of modern moral discourse, or even social observation? For this claim is entirely empty and devoid of explanatory power. We gain nothing by speculating that a capacity for genocide lies within our evolutionary heritage. We already know that we have such a capacity because human history provides so many examples of actualization.

An evolutionary speculation, to be useful, must suggest something we don't know already—if, for example, we learned that genocide has been biologically enjoined by certain genes, or even that a positive propensity, rather than a mere capacity, regulated our murderous potentiality. But the observational facts of human history speak against determination, and only for potentiality. Each case of genocide can be matched with numerous incidents of social benevolence; each murderous band can be paired with a pacific clan. Genocide gains greater prominence only for superior "news value"—and for devastating effectiveness (as the pacific clan disappears and murderers control the resulting media). But if both darkness and light lie within our capacities, and if both tendencies operate at high frequency in human history, then we learn nothing by speculating that either (or probably both) lie within our evolutionary, adaptive, Darwinian heritage. At the very most, biology might help us to delimit the environmental circumstances that tend to elicit one behavior rather than the other.

To cite the example most under current discussion in the "pop science" press, numerous books and articles preach to us that a new science of evolutionary psychology has discovered the biological basis of behavioral differences between sexes. Women produce only a few large eggs and must spend years of their lives growing embryos within their bodies and then nurturing the resulting babies. Men, on the other hand, produce millions of tiny sperm each time, and need invest nothing more in a potential offspring than the effort of an ejaculation. Therefore, the argument continues, in the great Darwinian quest for passing more genes to future generations, women will behave in ways that encourage male investment after impregnation (protection, feeding, economic wealth, and subsequent child care), whereas men would rather wander right off in search of other mates in a never-ending quest for maximal genetic spread. From this basic dichotomy of evolutionary purpose, all else in the lexicon of pop psychology follows. We now know why men rape, lust for power, dominate politics, have affairs, and abandon families with young children—and why women act coy, love to nurture children, and preferentially enter the caring professions.

Perhaps I have caricatured this position—but I don't think so, having read so many articles of support. In fact, I don't even think that the basic argument is wrong. Such differences in behavioral strategy do make Darwinian sense in the light of structural disparity between male and female reproduction. But the attributions could not be more deeply erroneous for the same reasons noted above in discussing the fallacy of biological explanations for genocide. Men are not programmed by genes to maximize matings, or women devoted to monogamy on the same basis. We can only speak of capacities, not of requirements or even determining propensities. Therefore, our biology does not make us do it. Moreover, what we share in common genetics can easily overwhelm what men and women might tend to do differently. Any man who has fiercely loved his little child—including most fathers, I trust (and I happen to be writing this essay on Father's Day)—knows that no siren song from distinctive genes or hormones can overcome this drive for nurturing behavior shared with the child's mother.

Finally, when we note the crucial differences in fundamental pattern and causation between biological evolution and cultural change—and when we recognize that everything distinctive about the cultural style enjoins flexibility rather than determination—we can understand even more generally why a cultural phenomenon like genocide (despite any underlying biological capacity for such action) cannot be explained in evolutionary terms. As the fundamental difference in pattern, biological evolution is a topological tree—a process of separation and divergence. A new species, arising as an independent lineage, acquires genetic distinction from all other lineages forever, and must evolve on its own path. Cultural change, on the other hand, is virtually defined by possibilities of amalgamation among different traditions—as Marco Polo brings pasta from China and I speak English as a "native" tongue. Our distinctive flexibilities arise from this constant interweaving.

As the fundamental difference in causation, biological evolution is Mendelian. Organisms can only pass their genes, not the heritage of their efforts, as physical contributions to future generations. But cultural change is Lamarckian, as we transmit the fruits of our acquired wisdom and inventiveness directly to future generations in the form of books, instruction manuals, tools, and buildings. Again, this Lamarckian style grants to cultural change a speed, a lability, and a flexibility that Darwinian evolution cannot muster.

In 1525, thousands of German peasants were slaughtered (with Luther's approbation), and Michelangelo worked on the Medici Chapel. In 1618, the upper windows of Prague disgorged a few men, and Rubens painted some mighty canvases. The Cathedral of Canterbury is both the site of Becket's murder and the finest Gothic building in England. Both sides of this dichotomy represent our common, evolved humanity; which, ultimately, shall we choose? As for the potential path of genocide and destruction, let us take this stand. It need not be. We can do otherwise.

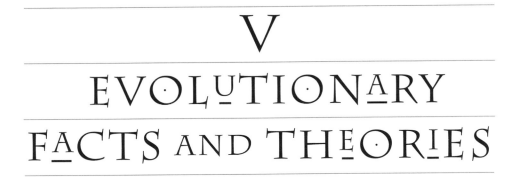

V

EVOLUTIONARY
FACTS AND THEORIES

14

NON-OVERLAPPING

MAGISTERIA

INCONGRUOUS PLACES OFTEN INSPIRE ANOMALOUS STORIES. IN EARLY 1984, I spent several nights at the Vatican housed in a hotel built for itinerant priests. While pondering over such puzzling issues as the intended function of the bidet in each bathroom, and hungering for something more than plum jam on my breakfast rolls (why did the basket only contain hundreds of identical plum packets and not a one of, say, strawberry?), I encountered yet another among the innumerable issues of contrasting cultures that can make life so expansive and interesting. Our crowd (present in Rome to attend a meeting on nuclear winter, sponsored by the Pontifical Academy of Sciences) shared the hotel with a group of French and Italian Jesuit priests who were also professional scientists. One day at lunch, the priests called me over to their table to pose a problem that had been troubling them. What, they wanted to know, was going on in America with all this talk about "scientific creationism"? One of the priests asked me: "Is evolution really in some kind of trouble; and, if so, what could such trouble be? I have always been taught that no doctrinal conflict exists between evolution and Catholic faith, and the evidence for evolution seems both utterly satisfying and entirely overwhelming. Have I missed something?"

A lively pastiche of French, Italian, and English conversation then ensued for half an hour or so, but the priests all seemed reassured by my general answer—"Evolution has encountered no intellectual trouble; no new arguments have been offered. Creationism is a home-grown phenomenon of American sociocultural history—a splinter movement (unfortunately rather more of a beam these days) of Protestant fundamentalists who believe that every word of the Bible must be literally true, whatever such a claim might mean." We all left satisfied, but I certainly felt bemused by the anomaly of my role as a Jewish agnostic, trying to reassure a group of priests that evolution remained both true and entirely consistent with religious belief.

Another story in the same mold: I am often asked whether I ever encounter creationism as a live issue among my Harvard undergraduate students. I reply that only once, in thirty years of teaching, did I experience such an incident. A very sincere and serious freshman student came to my office with a question that had clearly been troubling him deeply. He said to me, "I am a devout Christian and have never had any reason to doubt evolution, an idea that seems both exciting and well documented. But my roommate, a proselytizing evangelical, has been insisting with enormous vigor that I cannot be both a real Christian and an evolutionist. So tell me, can a person believe both in God and in evolution?" Again, I gulped hard, did my intellectual duty, and reassured him that evolution was both true and entirely compatible with Christian belief—a position that I hold sincerely, but still an odd situation for a Jewish agnostic.

These two stories illustrate a cardinal point, frequently unrecognized but absolutely central to any understanding of the status and impact of the politically potent, fundamentalist doctrine known by its self-proclaimed oxymoron as "scientific creationism"—the claim that the Bible is literally true, that all organisms were created during six days of twenty-four hours, that the earth is only a few thousand years old, and that evolution must therefore be false. Creationism does not pit science against religion (as my opening stories indicate), for no such conflict exists. Creationism does not raise any unsettled intellectual issues about the nature of biology or the history of life. Creationism is a local and parochial movement, powerful only in the United States among Western nations, and prevalent only among the few sectors of American Protestantism that choose to read the Bible as an inerrant document, literally true in every jot and tittle.

I do not doubt that one could find an occasional nun who would prefer to teach creationism in her parochial school biology class, or an occasional rabbi who does the same in his yeshiva, but creationism based on biblical literalism makes little sense either to Catholics or Jews, for neither religion maintains any extensive tradition for reading the Bible as literal truth, rather than illuminating literature based partly on metaphor and allegory (essential components of all good writing), and demanding interpretation for proper understanding. Most Protestant groups, of course, take the same position—the fundamentalist fringe notwithstanding.

The argument that I have just outlined by personal stories and general statements represents the standard attitude of all major Western religions (and of Western science) today. (I cannot, through ignorance, speak of Eastern religions, though I suspect that the same position would prevail in most cases.) The *lack of conflict* between science and religion arises from a *lack of overlap* between their respective domains of professional expertise—science in the empirical constitution of the universe, and religion in the search for proper ethical values and the spiritual meaning of our lives. The attainment of wisdom in a full life requires extensive attention to both domains—for a great book tells us both that the truth can make us free, and that we will live in optimal harmony with our fellows when we learn to do justly, love mercy, and walk humbly.

In the context of this "standard" position, I was enormously puzzled by a statement issued by Pope John Paul II on October 22, 1996, to the Pontifical Academy of Sciences, the same body that had sponsored my earlier trip to the Vatican. In this document, titled "Truth Cannot Contradict Truth," the Pope defended both the evidence for evolution and the consistency of the theory with Catholic religious doctrine. Newspapers throughout the world responded with front-page headlines, as in *The New York Times* for October 25: "Pope Bolsters Church's Support for Scientific View of Evolution."

Now I know about "slow news days," and I do allow that nothing else was strongly competing for headlines at that particular moment. Still, I couldn't help feeling immensely puzzled by all the attention paid to the Pope's statement (while being wryly pleased, of course, for we need all the good press we can get, especially from respected outside sources). The Catholic Church does not oppose evolution, and has no reason to do so. Why had the Pope issued such a statement at all? And why had the press responded with an orgy of worldwide front-page coverage?

I could only conclude at first, and wrongly as I soon learned, that journalists throughout the world must deeply misunderstand the relationship between science and religion, and must therefore be elevating a minor papal comment to unwarranted notice. Perhaps most people really do think that a war exists between science and religion, and that evolution cannot be squared with a belief in God. In such a context, a papal admission of evolution's legitimate status might be regarded as major news indeed—a sort of modern equivalent for a story that never happened, but would have made the biggest journalistic splash of 1640: Pope Urban VIII releases his most famous prisoner from house arrest and humbly apologizes: "Sorry, Signor Galileo . . . the sun, er, is central."

But I then discovered that such prominent coverage of papal satisfaction with evolution had not been an error of non-Catholic anglophone journalists. The Vatican itself had issued the statement as a major news release. And Italian newspapers had featured, if anything, even bigger headlines and longer stories. The conservative *Il Giornale,* for example, shouted from its masthead: "Pope Says We May Descend from Monkeys."

Clearly, I was out to lunch; something novel or surprising must lurk within the papal statement, but what could be causing all the fuss?—especially given the accuracy of my primary impression (as I later verified) that the Catholic Church values scientific study, views science as no threat to religion in general or Catholic doctrine in particular, and has long accepted both the legitimacy of evolution as a field of study and the potential harmony of evolutionary conclusions with Catholic faith.

As a former constituent of Tip O'Neill, I certainly know that "all politics is local"—and that the Vatican undoubtedly has its own internal reasons, quite opaque to me, for announcing papal support of evolution in a major statement. Still, I reasoned that I must be missing some important key, and I felt quite frustrated. I then remembered the primary rule of intellectual life: When puzzled, it never hurts to read the primary documents—a rather simple and self-evident principle that has, nonetheless, completely disappeared from large sectors of the American experience.

I knew that Pope Pius XII (not one of my favorite figures in twentieth-century history, to say the least) had made the primary statement in a 1950 encyclical entitled *Humani Generis*. I knew the main thrust of his message: Catholics could believe whatever science determined about the evolution of the human body, so long as they accepted that, at some time of his choosing, God had infused the soul into such a creature. I also knew that I had no problem with this argument—for, whatever my private beliefs about souls, science cannot touch such a subject and therefore cannot be threatened by any theological position on such a legitimately and intrinsically religious issue. Pope Pius XII, in other words, had properly acknowledged and respected the separate domains of science and theology. Thus, I found myself in total agreement with *Humani Generis*—but I had never read the document in full (not much of an impediment to stating an opinion these days).

I quickly got the relevant writings from, of all places, the Internet. (The Pope is prominently on line, but a luddite like me is not. So I got a cyberwise associate to dredge up the documents. I do love the fracture of stereotypes implied by finding religion so hep and a scientist so square.) Having now read in full both Pope Pius's *Humani Generis* of 1950 and Pope John

Paul's proclamation of October 1996, I finally understand why the recent statement seems so new, revealing, and worthy of all those headlines. And the message could not be more welcome for evolutionists, and friends of both science and religion.

The text of *Humani Generis* focuses on the *Magisterium* (or Teaching Authority) of the Church—a word derived not from any concept of majesty or unquestionable awe, but from the different notion of teaching, for *magister* means "teacher" in Latin. We may, I think, adopt this word and concept to express the central point of this essay and the principled resolution of supposed "conflict" or "warfare" between science and religion. No such conflict should exist because each subject has a legitimate magisterium, or domain of teaching authority—and these magisteria do not overlap (the principle that I would like to designate as NOMA, or "non-overlapping magisteria"). The net of science covers the empirical realm: what is the universe made of (fact) and why does it work this way (theory). The net of religion extends over questions of moral meaning and value. These two magisteria do not overlap, nor do they encompass all inquiry (consider, for starters, the magisterium of art and the meaning of beauty). To cite the usual clichés, we get the age of rocks, and religion retains the rock of ages; we study how the heavens go, and they determine how to go to heaven.

This resolution might remain entirely neat and clean if the non-overlapping magisteria of science and religion stood far apart, separated by an extensive no-man's-land. But, in fact, the two magisteria bump right up against each other, interdigitating in wondrously complex ways along their joint border. Many of our deepest questions call upon aspects of both magisteria for different parts of a full answer—and the sorting of legitimate domains can become quite complex and difficult. To cite just two broad questions involving both evolutionary facts and moral arguments: Since evolution made us the only earthly creatures with advanced consciousness, what responsibilities are so entailed for our relations with other species? What do our genealogical ties with other organisms imply about the meaning of human life?

Pius XII's *Humani Generis* (1950), a highly traditionalist document written by a deeply conservative man, faces all the "isms" and cynicisms that rode the wake of World War II and informed the struggle to rebuild human decency from the ashes of the Holocaust. The encyclical bears the subtitle "concerning some false opinions which threaten to undermine the foundations of Catholic doctrine," and begins with a statement of embattlement:

> Disagreement and error among men on moral and religious matters have always been a cause of profound sorrow to all good men, but above all to the true and loyal sons of the Church, especially today, when we see the principles of Christian culture being attacked on all sides.

Pius lashes out, in turn, at various external enemies of the Church: pantheism, existentialism, dialectical materialism, historicism, and, of course and preeminently, communism. He then notes with sadness that some well-meaning folks within the Church have fallen into a dangerous relativism— "a theological pacifism and egalitarianism, in which all points of view become equally valid"—in order to include those who yearn for the embrace of Christian religion, but do not wish to accept the particularly Catholic magisterium.

Speaking as a conservative's conservative, Pius laments:

> Novelties of this kind have already borne their deadly fruit in almost all branches of theology . . . Some question whether angels are personal beings, and whether matter and spirit differ essentially . . . Some even say that the doctrine of Transubstantiation, based on an antiquated philosophic notion of substance, should be so modified that the Real Presence of Christ in the Holy Eucharist be reduced to a kind of symbolism.

Pius first mentions evolution to decry a misuse by overextension among zealous supporters of the anathematized "isms":

> Some imprudently and indiscreetly hold that evolution . . . explains
> the origin of all things . . . Communists gladly subscribe to this opin-
> ion so that, when the souls of men have been deprived of every idea
> of a personal God, they may the more efficaciously defend and
> propagate their dialectical materialism.

Pius presents his major statement on evolution near the end of the encyclical, in paragraphs 35 through 37. He accepts the standard model of non-overlapping magisteria (NOMA) and begins by acknowledging that evolution lies in a difficult area where the domains press hard against each other. "It remains for Us now to speak about those questions which, although they pertain to the positive sciences, are nevertheless more or less connected with the truths of the Christian faith."[1]

Pius then writes the well-known words that permit Catholics to entertain the evolution of the human body (a factual issue under the magisterium of science), so long as they accept the divine creation and infusion of the soul (a theological notion under the magisterium of religion).

> The Teaching Authority of the Church does not forbid that, in con-
> formity with the present state of human sciences and sacred theol-
> ogy, research and discussions, on the part of men experienced in
> both fields, take place with regard to the doctrine of evolution, in as
> far as it inquires into the origin of the human body as coming from

---

1. Interestingly, the main thrust of these paragraphs does not address evolution in general, but lies in refuting a doctrine that Pius calls "polygenism," or the notion of human ancestry from multiple parents—for he regards such an idea as incompatible with the doctrine of original sin "which proceeds from a sin actually committed by an individual Adam and which, through generation, is passed on to all and is in everyone as his own." In this one instance, Pius may be transgressing the NOMA principle—but I cannot judge, for I do not understand the details of Catholic theology and therefore do not know how symbolically such a statement may be read. If Pius is arguing that we cannot entertain a theory about derivation of all modern humans from an ancestral population rather than through an ancestral individual (a potential fact) because such an idea would question the doctrine of original sin (a theological construct), then I would declare him out of line for letting the magisterium of religion dictate a conclusion within the magisterium of science.

pre-existent and living matter—for the Catholic faith obliges us to hold that souls are immediately created by God.

I had, up to here, found nothing surprising in *Humani Generis,* and nothing to relieve my puzzlement about the novelty of Pope John Paul's recent statement. But I read further and realized that Pius had said more about evolution, something I had never seen quoted, and something that made John Paul's statement most interesting indeed. In short, Pius forcefully proclaimed that while evolution may be legitimate in principle, the theory, in fact, had not been proven and might well be entirely wrong. One gets the strong impression, moreover, that Pius was rooting pretty hard for a verdict of falsity.

Continuing directly from the last quotation, Pius advises us about the proper study of evolution:

> However, this must be done in such a way that the reasons for both opinions, that is, those favorable and those unfavorable to evolution, be weighed and judged with the necessary seriousness, moderation and measure . . . Some however, rashly transgress this liberty of discussion, when they act as if the origin of the human body from pre existing and living matter were already completely certain and proved by the facts which have been discovered up to now and by reasoning on those facts, and as if there were nothing in the sources of divine revelation which demands the greatest moderation and caution in this question.

To summarize, Pius generally accepts the NOMA principle of non-overlapping magisteria in permitting Catholics to entertain the hypothesis of evolution for the human body so long as they accept the divine infusion of the soul. But he then offers some (holy) fatherly advice to scientists about the status of evolution as a scientific concept: the idea is not yet proven, and you all need to be especially cautious because evolution raises many troubling issues right on the border of my magisterium. One may read this second theme in two rather different ways: either as a gratuitous incursion into a different magisterium, or as a helpful perspective from an intelligent and concerned outsider. As a man of goodwill, and in the interest of conciliation, I am content to embrace the latter reading.

In any case, this rarely quoted second claim (that evolution remains both unproven and a bit dangerous)—and not the familiar first argument for the NOMA principle (that Catholics may accept the evolution of the body so long as they embrace the creation of the soul)—defines the novelty and the interest of John Paul's recent statement.

John Paul begins by summarizing Pius's older encyclical of 1950, and particularly by reaffirming the NOMA principle—nothing new here, and no cause for extended publicity:

> In his encyclical "Humani Generis" (1950), my predecessor Pius XII had already stated that there was no opposition between evolution and the doctrine of the faith about man and his vocation.

To emphasize the power of NOMA, John Paul poses a potential problem and a sound resolution: How can we possibly reconcile science's claim for physical continuity in human evolution with Catholicism's insistence that the soul must enter at a moment of divine infusion?

> With man, then, we find ourselves in the presence of an ontological difference, an ontological leap, one could say. However, does not the posing of such ontological discontinuity run counter to that physical continuity which seems to be the main thread of research into evolution in the field of physics and chemistry? Consideration of the method used in the various branches of knowledge makes it possible to reconcile two points of view which would seem irreconcilable. The sciences of observation describe and measure the multiple manifestations of life with increasing precision and correlate them with the time line. The moment of transition to the spiritual cannot be the object of this kind of observation.

The novelty and news value of John Paul's statement lies, rather, in his profound revision of Pius's second and rarely quoted claim that evolution, while conceivable in principle and reconcilable with religion, can cite little persuasive evidence in support, and may well be false. John Paul states—and I can only say amen, and thanks for noticing—that the half century between Pius surveying the ruins of World War II and his own pontificate heralding the dawn of a new millennium has witnessed such a growth of data, and such a refinement of theory, that evolution can no longer be doubted by people of goodwill and keen intellect:

> Pius XII added . . . that this opinion [evolution] should not be adopted as though it were a certain, proven doctrine . . . Today, almost half a century after the publication of the encyclical, new knowledge has led to the recognition of the theory of evolution as more than a hypothesis.[2] It is indeed remarkable that this theory has been progressively accepted by researchers, following a series of

2. This passage, here correctly translated, provides a fascinating example of the subtleties and inherent ambiguities in rendering one language into another. Translation may be the most difficult of all arts, and meanings have been reversed (and wars fought) for perfectly understandable reasons. The Pope originally issued his statement in French, where this phrase read ". . . *de nouvelles connaissances conduisent à reconnaitre dans la théorie de l'évolution plus qu'une hypothèse.*" *L'Osservatore Romano,* the official Vatican newspaper, translated this passage as: "new knowledge has led to the recognition of more than one hypothesis in the theory of evolution." This version (obviously, given the official Vatican source) then appeared in all English commentaries, including the original version of this essay.

I included this original translation, but I was profoundly puzzled. Why should the Pope be speaking of *several* hypotheses within the framework of evolutionary theory? But I had no means to resolve my confusion, so I assumed that the Pope had probably fallen under the false impression (a fairly common misconception) that, although evolution had been documented beyond reasonable doubt, natural selection had fallen under suspicion as a primary mechanism, while other alternatives had risen to prominence.

Other theologians and scientists were equally puzzled, leading to inquiries and a resolution of the problem as an error in translation (as many of us would have realized right away if we had seen the original French, or even known that the document had been issued in French). The problem lies with ambiguity in double meaning for the indefinite article in French—where *un* (feminine *une*) can mean either "a" or "one." Clearly, the Pope had meant that the theory of evolution had now become strong enough to rank as "more than *a* hypothesis" (*plus qu'une*

discoveries in various fields of knowledge. The convergence, neither sought nor fabricated, of the results of work that was conducted independently is in itself a significant argument in favor of the theory.

In conclusion, Pius had grudgingly admitted evolution as a legitimate hypothesis that he regarded as only tentatively supported and potentially (as he clearly hoped) untrue. John Paul, nearly fifty years later, reaffirms the legitimacy of evolution under the NOMA principle—no news here—but then adds that additional data and theory have placed the factuality of evolution beyond reasonable doubt. Sincere Christians must now accept evolution not merely as a plausible possibility, but also as an effectively proven fact. In other words, official Catholic opinion on evolution has moved from "say it ain't so, but we can deal with it if we have to" (Pius's grudging view of 1950) to John Paul's entirely welcoming "it has been proven true; we always celebrate nature s factuality, and we look forward to interesting discussions of theological implications." I happily endorse this turn of events as gospel—literally good news. I may represent the magisterium of science, but I welcome the support of a primary leader from the other major magisterium of our complex lives. And I recall the wisdom of King Solomon: "As cold waters to a thirsty soul, so is good news from a far country" (Proverbs 25:25).

Just as religion must bear the cross of its hard-liners, I have some scien-

---

hypothèse), but the Vatican originally read *une* as "one" and gave the almost opposite rendition: "more than *one* hypothesis." *Caveat emptor.*

I thank about a dozen correspondents for pointing out this error, and the Vatican's acknowledgment, to me. I am especially grateful to Boyce Rensberger, one of America's most astute journalists on evolutionary subjects, and David M. Byers, executive director of the National Conference of Catholic Bishops' Committee on Science and Human Values. Byers affirms the NOMA principle by writing to me: "Thank you for your recent article . . . It admirably captures the relationship between science and religion that the Catholic Bishops' Committee works to promote and to realize. The text of the October 1996 papal statement from which you were working contains a mistranslation of a key phrase; the correct translation supports your thesis with even greater force."

tific colleagues, including a few in prominent enough positions to wield influence by their writings, who view this rapprochement of the separate magisteria with dismay. To colleagues like me—agnostic scientists who welcome and celebrate the rapprochement, especially the Pope's latest statement—they say, "C'mon, be honest; you know that religion is addlepated, superstitious, old-fashioned BS. You're only making those welcoming noises because religion is so powerful, and we need to be diplomatic in order to buy public support for science." I do not think that many scientists hold this view, but such a position fills me with dismay—and I therefore end this essay with a personal statement about religion, as a testimony to what I regard as a virtual consensus among thoughtful scientists (who support the NOMA principle as firmly as the Pope does).

I am not, personally, a believer or a religious man in any sense of institutional commitment or practice. But I have great respect for religion, and the subject has always fascinated me, beyond almost all others (with a few exceptions, like evolution and paleontology). Much of this fascination lies in the stunning historical paradox that organized religion has fostered, throughout Western history, both the most unspeakable horrors and the most heartrending examples of human goodness in the face of personal danger. (The evil, I believe, lies in an occasional confluence of religion with secular power. The Catholic Church has sponsored its share of horrors, from Inquisitions to liquidations—but only because this institution held great secular power during so much of Western history. When my folks held such sway, more briefly and in Old Testament times, we committed similar atrocities with the same rationales.)

I believe, with all my heart, in a respectful, even loving, concordat between our magisteria—the NOMA concept. NOMA represents a principled position on moral and intellectual grounds, not a merely diplomatic solution. NOMA also cuts both ways. If religion can no longer dictate the nature of factual conclusions residing properly within the magisterium of science, then scientists cannot claim higher insight into moral truth from any superior knowledge of the world's empirical constitution. This mutual humility leads to important practical consequences in a world of such diverse passions.

Religion is too important for too many people to permit any dismissal or denigration of the comfort still sought by many folks from theology. I may, for example, privately suspect that papal insistence on divine infusion of the soul represents a sop to our fears, a device for maintaining a belief in human superiority within an evolutionary world offering no privileged position to any creature. But I also know that the subject of souls lies outside the magisterium of science. My world cannot prove or disprove such a notion, and the concept of souls cannot threaten or impact my domain. Moreover, while I cannot personally accept the Catholic view of souls, I surely honor the metaphorical value of such a concept both for grounding moral discussion, and for expressing what we most value about human potentiality: our decency, our care, and all the ethical and intellectual struggles that the evolution of consciousness imposed upon us.

As a moral position (and therefore not as a deduction from my knowledge of nature's factuality), I prefer the "cold bath" theory that nature can be truly "cruel" and "indifferent"—in the utterly inappropriate terms of our ethical discourse—because nature does not exist for us, didn't know we were coming (we are, after all, interlopers of the latest geological moment), and doesn't give a damn about us (speaking metaphorically). I regard such a position as liberating, not depressing, because we then gain the capacity to conduct moral discourse—and nothing could be more important—in our own terms, free from the delusion that we might read moral truth passively from nature's factuality.

But I recognize that such a position frightens many people, and that a more spiritual view of nature retains broad appeal (acknowledging the factuality of evolution, but still seeking some intrinsic meaning in human terms, and from the magisterium of religion). I do appreciate, for example, the struggles of a man who wrote to *The New York Times* on November 3, 1996, to declare both his pain and his endorsement of John Paul's statement:

Pope John Paul II's acceptance of evolution touches the doubt in my heart. The problem of pain and suffering in a world created by a God who is all love and light is hard enough to bear, even if one is a creationist. But at least a creationist can say that the original creation, coming from the hand of God, was good, harmonious, innocent and gentle. What can one say about evolution, even a spiritual theory of evolution? Pain and suffering, mindless cruelty and terror are its means of creation. Evolution's engine is the grinding of predatory teeth upon the screaming, living flesh and bones of prey . . . If evolution be true, my faith has rougher seas to sail.

I don't agree with this man, but we could have a terrific argument. I would push the "cold bath" theory; he would (presumably) advocate the theme of inherent spiritual meaning in nature, however opaque the signal. But we would both be enlightened and filled with better understanding of these deep and ultimately unanswerable issues. Here, I believe, lies the greatest strength and necessity of NOMA, the non-overlapping magisteria of science and religion. NOMA permits—indeed enjoins—the prospect of respectful discourse, of constant input from both magisteria toward the common goal of wisdom. If human beings can lay claim to anything special, we evolved as the only creatures that must ponder and talk. Pope John Paul II would surely point out to me that his magisterium has always recognized this uniqueness, for John's gospel begins by stating *in principio erat verbum*—in the beginning was the word.

1 5

BOYLE'S LAW AND

DARWIN'S DETAILS

TWO SCENES FROM FLORENCE BEAUTIFULLY ILLUSTRATE THE POWER OF scientific revolutions to alter our view of the geometry of existence. A painting by the fifteenth-century artist Michellino hangs in the great cathedral of Santa Maria del Fiore. Titled *Dante e il suo poema* ("Dante and His Poem"—that is, *The Divine Comedy*), it shows the entire universe on a single canvas. The earth occupies the center, symbolized by the city of Florence, with Dante in the middle and Brunelleschi's magnificent dome for the cathedral to his left (an anachronism, to be sure, for Dante died in 1321 and Brunelleschi raised the great dome a century later). At Dante's right, the souls of the damned move downward to the inferno, while those destined for ultimate salvation slowly mount the spire of purgatory. The seven semicircles at the top represent the seven planets of Ptolemy's earth-centered system (the five visible planets plus the sun and the moon). The farthest realm of the fixed stars occupies the upper corners.

If we take a short walk to the Franciscan church of Santa Croce, we find the tomb of Galileo. He looks upward toward his expanded heavens and holds a telescope in his right hand. His left hand envelops the small and insignificant sphere of the earth. In two centuries (Galileo died in 1642), the earth had been displaced from a large and ruling centrality in a limited and subservient universe to peripheral status as a little hunk of stone suspended in the midst of inconceivable vastness.

In a famous statement, Sigmund Freud argued that scientific revolutions reach completion not when people accept the physical reconstruction of reality thus implied, but when they also own the consequences of this radically revised universe for a demoted view of human status. Freud claimed that all important scientific revolutions share the ironic property of deposing humans from one pedestal after another of previous self-assurance about our exalted cosmic status. Therefore, all great revolutions smash pedestals— and inspire resistance for the obvious reason that we accept such demotions only begrudgingly. Freud identified two revolutions as paramount— Copernicus and Galileo on the nature of the heavens, and Darwin on the status of life. Unfortunately, Darwin's revolution remains incomplete to this day because we spin-doctor the results of evolution to preserve our pedestal of arrogance by misreading the process as a predictable accumulation of improvements, leading sensibly to the late appearance of human intelligence as a culmination.

Although we have yet to make our peace with Darwin, the first revolution of cosmic realignment passed quickly into public acceptance. In 1633, Galileo appeared before the Inquisition in Rome, where, under threat of torture and death, he officially abjured his belief in the sun-centered Copernican system. He spent the rest of his life under house arrest on his estate at Arcetri, near Florence, where he died in 1642. In the same year, Robert Boyle, then a wealthy teenager on his grand tour of Europe, but soon to become a great physicist and chemist in his own right, visited Florence and read Galileo's *Dialogue on the Two Chief World Systems*—just as the master lay dying nearby in Arcetri.

Boyle's Law and Darwin's Details

In 1688, as an elderly man himself, Robert Boyle wrote a famous treatise on science and religion titled *A disquisition about the final causes of natural things: wherein it is inquir'd, whether, and (if at all) with what cautions, a naturalist should admit them?* In this work, just two generations after Galileo's death, Boyle demonstrates that the pedestal-smashing implications of Copernican cosmology had already been articulated and accepted, thus completing the first revolution in Freud's crucial sense. (I regard this timing as important because one might claim that Galileo triumphed, while Darwin remains in limbo, simply because pedestal-smashing takes centuries and Galileo had a two-hundred-year head start. But if the pedestal crumbled during Galileo's own century, then we have had more than enough time for Darwin—and our failure to smash this second pedestal must record its greater durability in our unwilling psyches.)

Boyle asks whether the existence and regular motion of the sun and moon provide evidence of God's creative power and benevolent action. He begins by ridiculing those who would argue for the old geocentric system because God made everything for human benefit:

> I dare not imitate their boldness, that affirm, that the sun and moon, and all the stars, and other celestial bodies, were made solely for the use of man; . . . as when they argue, that the sun and other vast globes of light, ought to be in perpetual motion to shine upon the earth; because, they fancy, 'tis more convenient for man, that those distant bodies, than that the earth, which is his habitation, should be kept in motion.

Boyle then invokes the smashed pedestal more directly to claim that God would not create something so huge as the sun only to illuminate such a tiny and inconsequential body as the earth:

> But, considering things as mere naturalists, it seems not very likely, that a most Wise Agent should have made such vast bodies, as the

sun and the fixed stars, especially if we suppose them to move with that inconceivable rapidity that vulgar astronomers do and must assign them; only or chiefly to illuminate a little globe that without hyperbole is but a physical point in comparison of the immense spaces comprised under the name of heaven.

We would not expect Boyle, who (after all) wrote 150 years before Darwin, to assault the second pedestal or even to question the creationist view of life at all. Rather, I dedicate this essay to demonstrating that Boyle's particular view of natural religion provides a distinctively English insight into the historical traditions that make the Darwinian pedestal so impervious to demolition. I then show that Darwin's philosophical radicalism lies best exposed when we view the theory of natural selection as a direct and purposeful assault upon Boyle's natural theology.

Robert Boyle (1627–1691), son of the first Earl of Cork, belonged to a noble and wealthy Anglo-Irish family. After studying at Eton and living abroad for several years, Boyle spent his most scientifically productive decade in Oxford (1656–1668), where he constructed an air pump and performed his major experiments on the properties of gases. (His most famous result, Boyle's Law, states that, at a constant temperature, the pressure of a given quantity of gas varies inversely to its volume.) In his major work, the *Sceptical Chemist* (1661), Boyle attacked the Aristotelian theory of four elements (earth, air, fire, and water), and developed an important corpuscular theory of matter. (He did not postulate different kinds of basic elements, as later validated and established in the periodic table, but rather argued that properties of different substances arose from variation in the motion and organization of primary particles.)

Boyle moved permanently to London in 1668, where he continued his organizational work as a founder of the Royal Society (still Britain's leading scientific establishment), and labored for many other causes close to his heart. (He was, for example, the governor of the Society for the Propagation of the Gospel in New England.)

Boyle's Law and Darwin's Details

In science, Boyle's main reputation rested upon his stern defense of mechanism and his abjuration of Aristotelian forms and essences. The *Dictionary of Scientific Biography* describes his fundamental philosophy in these terms:

> [Boyle was] a profound believer in the need to establish an empirically based, mechanistic theory of matter and in the possibility of establishing a scientific, rational, theoretical chemistry . . . Boyle was long remembered as "the restorer of the mechanical philosophy" in England . . . What mattered most to him was destroying all Aristotelian forms of qualities . . . and substituting for them rational, mechanical explanation in terms of what he called "those two grand and most catholic principles, matter and motion."

But Boyle matched his devotion to science with another controlling and passionate interest—his orthodox Protestant beliefs and his unflagging commitment to the cause of religion. Of all the scientists in Newton's orbit, Boyle was the most conventionally and sincerely devout. Moreover, Boyle did not consider religion as a merely private matter. He wrote as much about theology as about science, and he composed several treatises on the potentially harmonious relationship between these two disciplines, including the work analyzed in this essay.

Such a statement may seem contradictory if we accept the false, but commonly held, view that all religion must be inherently mystical, while the mechanistic components of science must be antithetical to such a notion of higher reality. But Boyle's view of God, widely shared by Newton and most of his scientific contemporaries, neatly married mechanism and religion into a coherent system that granted higher status to both sides. Boyle's God is a masterful mechanical clock-winder who created the universe with all natural laws so perfectly tuned and contrived at the outset that the entire course of future history could unfold without further miraculous intervention (though neither Boyle nor Newton wanted to constrain God to His initial decisions, and certainly granted Him the right to interpose a miracle or two now and then, whenever His ineffable wisdom so decreed). Boyle wrote:

> The most wise and powerful Author of Nature, whose piercing sight is able to penetrate the whole universe, and survey all the parts of it at once, did at the beginning of things, frame things corporeal into such a system, and settled among them such laws of motion, as he judged suitable to the ends he proposed to himself in making the world. And by virtue of his vast and boundless intellect that he at first employed, he was able not only to see the present state of things he had made, but to foresee all the effects . . . Nor is this doctrine inconsistent with the belief of any true miracle; for, it supposes the ordinary and settled course of nature to be maintained, without at all denying, that the most free and powerful Author of Nature is able, whenever he thinks fit, to suspend, alter, or contradict those laws of motion, which he alone at first established.

Since God's invariant laws can be discovered and studied by science, and since divine omnipotence lies best exposed in these regularities of nature, God's glory can be apprehended empirically, thus making science a hand-maid to religion, and not an adversary at all.

Boyle's 1688 *Disquisition About the Final Causes of Natural Things,* though rarely consulted today (and undoubtedly unknown to nearly all practicing scientists), stands as the classic statement of this English approach to natural theology. Boyle's book began a tradition that culminated in one of the most influential books of the nineteenth century, William Paley's *Natural Theology* of 1802, and collapsed with Darwin's *Origin of Species* in 1859. As the centerpiece of this tradition, Boyle and his colleagues proposed and developed the so-called argument from design—the attempt to identify final causes in nature as proofs both for God's existence and for His attributes of ultimate power and unceasing benevolence. (Paley subtitled his work

Evidences of the Existence and Attributes of the Deity, Collected from the Appearances of Nature.)

To appreciate the power (and ultimate fallacy) of this argument, we must recover some forgotten terminology to explicate the notion of "final cause" in the light of Aristotle's celebrated analysis. (Boyle cared little for Aristotle's physics, but he followed the master's traditional explication of the categories of causality.) Aristotle proposed that causality included four separate components, as illustrated in the classic "parable of the house." The stuff of construction counts as a "material cause," for the house will be different (as the three little pigs discovered) if the building be made of straw, wood, or stone. The mason who actually puts the pieces together works as an "efficient cause." The blueprint only represents a plan on paper and does nothing to build the house directly. But without a plan, you never get beyond a pile of stones, so the blueprint counts as a "formal cause." Lastly, without an intended purpose, no one would bother to build at all, so the owner's desire to live in the house counts as a "final cause" (not "final" in the temporal sense of coming last, but in the literal Latin meaning of an end, or a purpose).

In one of the most striking changes in the definition of science between Boyle's day and our own, our concept of the meaning of causality has shifted in a fundamental way. One change is only terminological and therefore less important. We would still acknowledge the vital character of material and formal factors, but we no longer choose to label them as "causes." As the fundamental change, the mechanical revolution unleashed by Boyle and his generation succeeded so thoroughly that the actual building and manipulation of things, called "efficient" causes by Aristotle, became enshrined as the only acceptable definition of causality. Meanwhile, and in consequence, the notion of purpose, or final cause, was banished from science. We no longer believe that inorganic objects have intended purposes, defined either in human or in any other terms. As for organisms, we surely allow a notion of purpose in the vernacular sense that good designs have functions (yes, eyes

evolved for seeing)—but we now view such functions as products of the efficient cause of natural selection, and not as conscious intentions either of organisms themselves or of a creating deity.

But final causes remained orthodox (in science as well as religion) during Boyle's time, and he wrote his 1688 treatise to define the appropriate domain of final causes and to assess the evidence for their action. Interestingly, Boyle sets up the issue as a "Goldilocks" problem by identifying one class of objects as too little, another as too big, and a third as just right. In defining the good design of organisms as "just right," Boyle firmly linked the venerable notion of final causes to biology, and therefore rooted his natural defense of religion in the phenomena that Darwin's revolution would later identify as a product of ordinary efficient causality. Boyle's argument—that good organic design implies benevolent purpose in the cosmos—provides a comfort and appeal that we have not been able to relinquish. So we shore up the pedestal that Darwin should have smashed, and we spin-doctor our interpretation of evolution to view organic change as predictably purposeful (rather than fortuitously contingent), thereby converting Darwin's mechanism into a false argument for the same comfort that Boyle's God once provided.

Boyle begins his argument by stating that two schools of philosophical thought deny the existence of ascertainable final causes for opposite reasons—the Epicureans, who view material objects as constructed by chance, and the Cartesians, who regard God's wisdom as so inscrutable that mere earthly mortals could never discern his true purposes:

> Epicurus and most of his followers banish the consideration of the ends of things [final causes] because the world being, according to them, made by chance, no ends of anything can be supposed to have been intended. And on the contrary, Monsieur Des Cartes, and most of his followers, suppose all the ends of God in things corporeal to be so sublime, that 'twere presumption in man to think his reason can extend to discover them. So that, according to these opposite

sects, 'tis either impertinent for us to seek after final causes, or presumptuous to think we may find them.

Boyle then applies his Goldilocks approach to ask what class of natural objects might display final causes indicating creation by an omniscient and loving deity. In Momma Bear's category of "too little," Boyle places the inorganic objects on our earth—"inanimate in the sublunary world," in his terminology. Rocks and waters are so simple in composition that they might either be formed by chance (and therefore subject to the Epicurean objections against final causes) or built directly by nature's constant and simple laws. (God ordained the laws, to be sure, but objects assembled by physical forces under laws of nature, and not created by God, do not display God's purposes directly.) Boyle writes:

> As for inanimate bodies, as stones, metals, etc. . . . most of them are of such easy and unelaborate contextures, that it seems not absurd to think that various occursions and jostlings of the parts of the universal matter, may at one time or another have produced them, since we see in some chemical sublimations and crystalizations of mineral and metalline solutions, and some other phenomena, where the motions appear not to be particularly guided and directed by an intelligent Cause that bodies of various contextures as those are wont to be, may be produced.

In Poppa Bear's category of "too much," Boyle places the massive, inorganic objects of the cosmos—our sun, the planets, and the stars. They are so vast, so distant, so ineffable. God must have made them, but not for us (remember that Galileo and company had already smashed the first pedestal). These bodies, therefore, cannot display satisfactory final causes that might comfort or enlighten human beings. Stars and planets fall prey to the Cartesian claim that God's purposes are too arcane for human understanding. The stars extol God's greatness, but not his loving-kindness—and proper final causes must display both God's existence and his benevolence: "The Cartesian way of considering the world is very proper indeed to show the greatness of God's power, but not, like the way I plead for, to manifest that of his wisdom and beneficence."

What objects shall then occupy Baby Bear's category of "just right"— "the way I plead for," in Boyle's terminology. Boyle proposes animals and plants as proper evidence for final causes that prove God's existence and goodness. First, in contrast to Momma Bear's simplicity of inorganic objects, animals are sufficiently complex to require a direct creator:

> If we allow chance, or anything else, without the particular guid-ance of a wise and all-disposing cause, to make a finely shaped stone, or a metalline substance . . . there are others that require such a number of and concourse of conspiring causes, and such a contin-ued series of motions or operations, that 'tis utterly improbable, they should be produced without the superintendency of a rational agent, wise and powerful . . . I never saw any inanimate production of nature, or, as they speak, of chance, whose contrivance was com-parable to that of the meanest limb of the dispicablest animal.

Second, against Poppa Bear's ineffable grandeur of stars and planets, the parts of animals are familiar enough to reveal their purposes, and therefore to show their creator's intent: "I cannot but think," Boyle writes, "that the situations of the celestial bodies do not afford by far so clear and cogent arguments of the wisdom and design of the author of the world, as do the bodies of animals and plants."

I wish that I had space to explicate Boyle's detailed and clever (but also forced and ultimately invalid) arguments for optimal design and utility of every part and function of organisms. Instead, I will only discuss how this classic argument for God's existence and benevolence both builds and reinforces the pedestal that we seem so loath to smash in order to complete Darwin's revolution—for if the biosphere operates as a well-oiled machine of divine construction, and if purposes should be construed in terms of utility to humans (as the most perfect of all God's creatures), then natural theology affirms our domination and right to rule. One last time, in Boyle's words:

> The terraqueous globe and its productions . . . and especially the plants and animals 'tis furnished with, do . . . appear to have been designed for the use and benefit of man, who has therefore a right to employ as many of them as he is able to subdue . . . Therefore the kingly prophet had reason to exclaim: How manifold are thy works O Lord! How wisely hast thou made them all!

When Darwin set out, with conscious intent, to revolutionize human attitudes about the status and history of plants and animals, he did not deny Boyle's premise that organisms are well designed—and that excellence of anatomy and function establishes the primary problem for natural history to resolve. Darwin writes, in the preface to the *Origin of Species,* that evidence of taxonomy, embryology, paleontology, and biogeography would be sufficient to prove the operation of evolution, but that we could not be satisfied until we had explained "that perfection of structure and coadaptation which most justly excites our admiration."

But Darwin then turned Boyle and Paley upside down in accepting their premise (excellence of organic design), while inverting their explanation. Instead of a benevolent deity making organisms expressly for higher purposes (with human utility paramount among divine intentions), Darwin postulated a mechanistic process called natural selection (an efficient cause). Moreover, and most contrary to older traditions, Darwin's cause does not operate on such "higher" entities as species and ecosystems, but only on organisms struggling for personal reproductive success—and nothing else! The very features of nature that Boyle and Paley had read as proofs of God's existence and goodness—the excellent design of organisms and the harmony of ecosystems—became, for Darwin, side consequences or sequelae of a process without overarching purpose, and working directly only for the benefit of individual organisms!

Unlike Boyle, Darwin had no abiding interest in formal theology. But we must wonder what he thought about the wider implications of evolution and natural selection for human status. In other words, how much did Darwin explicitly desire to smash the pedestal that has prevented the completion of his revolution in Freud's sense? How far did he wish to go in undoing Boyle's traditional view of human domination (or at least superiority) in a sensibly constructed world?

Since Darwin did not write books about such philosophical questions, we must go to his private letters and jottings. One famous letter stands out as particularly revealing (and beautifully expressed)—a proof that Darwin aspired to revolutionary status in Freud's sense. Darwin's most famous American colleague, the Harvard botanist Asa Gray, read the *Origin of Species* (then hot off the presses) with both exhilaration and distress. In a heartfelt and deeply moving letter, Gray told Darwin that he could accept natural selection as an efficient cause of evolutionary change, but that, as a convinced theist, he could not abandon the conviction (however unprovable) that God must have revealed some higher purpose in constructing nature to work by such a principle. Darwin, in his wonderful reply of May 22, 1860, answered with compassion, but also with profound doubt about this traditional comfort:

> With respect to the theological view of the question. This is always painful to me. I am bewildered. I had no intention to write atheistically. But I own that I cannot see as plainly as others do, and as I should wish to do, evidence of design and beneficence on all sides of us. There seems to me too much misery in the world. I cannot persuade myself that a beneficent and omnipotent God would have designedly created the Ichneumonidae with the express intention of

their feeding within the living bodies of caterpillars, or that a cat should play with mice.

Darwin then penned his key line about design and intention in the history of life—in my view, one of the "great quotes" in the annals of Western thought:

> On the other hand, I cannot anyhow be contented to view this wonderful universe, and especially the nature of man, and to conclude that everything is the result of brute force. I am inclined to look at everything as resulting from designed laws, with the details, whether good or bad, left to the working out of what we may call chance.

We now reach the crucial point in explicating Darwin's view on this most fundamental of all questions. He can accept lawlike predictability, perhaps even with some underlying intent in some ill-defined theological sense, for background generalities. But Darwin also holds his hammer in the pedestal-smashing position for what he designates as "the details": they are left "to the working out of what we may call chance." By this careful choice of words, and by his examples, I am convinced that Darwin meant what we now call contingency (or unpredictability due to the extreme complexity of historical sequences), rather than chance in the dice-tossing sense. (This distinction could not be more crucial, because pure chance precludes any explanation of particulars, but contingency, while denying that predictions can be made with confidence at the outset, does assert the possibility of explanation after a particular history has unfolded. Contingency represents the historian's mode of knowability; pure chance denies that particulars can be explicated at all.)

We now come face to face with another Goldilocks problem. Darwin proposes a conventional realm of generalities and a revolutionary domain of particulars. But which factor dominates in the history of life? Do the particulars only represent a few insignificant bumps and pits on a ball that rolls according to fixed laws of motion, perhaps established with a purpose? Or do the particulars form mountains and gashes so high and deep that the ball's course must follow these dominating irregularities? Do the particulars lie in Momma Bear's little bed, or on Poppa Bear's king-sized mattress (sorry for the sexist implications of these categories, but I refuse to write politically correct bedtime stories—if only, and all the better, to acknowledge history's sad legacies).

Darwin's canny continuation of his argument to Gray indicates that he places the particulars in Poppa Bear's camp of "too much"—that is, too many and too influential to validate the traditional comforts of predictable human domination. He sneaks up on a ruling role for contingency with a series of three examples, the first two undeniable, the third more challenging, but eminently plausible once you accept the first two.

Example one: "The lightning kills a man, whether a good one or bad one, owing to the excessively complex action of natural laws." Fine. No arguments. The event was not random. The lightning struck a particular spot as an outcome of physical principles, but no one would say that the man happened to be in this spot by design. His death is contingent and unpredictable.

Example two: If we admit contingency for deaths, why not for births as well? "A child (who may turn out an idiot) is born by the action of even more complex laws." Again, if we understood embryology better, we would know (in a physical sense) why a child entered life with severe mental handicaps. But would we ever want to argue that a beneficent God intended such a particular, and tragic, result in establishing sensible principles of embryonic development? This particular outcome must be read as a contingency without moral meaning.

Example three: the evolutionary extension. Evolution is also a process of birth and death—of species and populations this time. If individual births (the retarded child) and deaths (the man killed by lightning) are contingent, then why not extend the same analysis to the birth and death of species as well—for species are the biological individuals of geological time scales.

And, since *Homo sapiens* is but one species among many, why should our birth (and potential death) be viewed as anything more than another contingency? "... and I can see no reason why a man, or other animal, may not have been aboriginally produced by other laws."

My high school drama teacher once told me that the most famous stage direction in English occurs in act III, scene iii of Shakespeare's *The Winter's Tale,* where the Bard writes, after Antigonus's long soliloquy: "Exit, pursued by a bear." Dare we hope that the false, harmful, and traditional comfort of evolutionary progress and human supremacy might finally make its exit—as Poppa Bear, strengthened by the dominating weight of contingency's realm, brings down his mallet upon the pedestal that Robert Boyle codified, William Paley enshrined, and Charles Darwin deprived of meaning?

16

THE TALLEST TALE

As a scholarly devotee of armchairs and ivory towers, I begin
with two strikingly similar legends about standing up at events designed for
sitting. In high culture's version, the audience rises at the opening chords of
the Hallelujah Chorus and remains standing throughout the piece. (Choral
singers—I am one—love the ritual, for we thereby obtain our only guaranteed standing ovation. The intermission after part two of Handel's *Messiah*
directly follows this great chorus.) In pop culture's primary example—the
seventh-inning stretch—fans by the tens of thousands stand before their
team comes to bat in the seventh inning of a baseball game. (The effect is
almost eerie. No one makes any announcement, and an unruly crowd
behaves, for this one moment, as an entity. Countless fathers have taken
advantage of this ritual by telling gullible children at their first ball game: "I
can make the entire audience stand at my command"—and then issuing the
appropriate order just before the predictable response. Has any kid ever
been tricked into obedience by believing that "father sees all"?)

Though names and places vary maximally, we tell exactly the same (undoubtedly false) story to explain each ritual. An English king (someone between George II and George IV, depending on your favored version) found Handel's majestic music so moving that he stood in honor—and audiences have done so ever since. An American president (William Howard Taft by consensus) got up at a ball game to stretch his legs, and everyone rose to honor the office.

I love these tales because, in more reasonable attributions of motive, they so beautifully embody a fundamental theme of historical explanation—that consequences of substantial import often arise from trivial triggers of entirely different intent. In other words, current utility bears no necessary relationship with historical origin. Who knows why good King George-the-whatever stood up? Maybe he thought the intermission had already come? Maybe he was bored, or wanted to go out for a smoke? As for Taft, he probably got up to leave early (some versions even recount the story this way). Has any president ever stayed for an entire game? But think of the aggregated consequences ever since—millions upon millions of people standing at the appointed time. Gazillions of joules in spent energy. All manner of secondarily accreting traditions, as people who never otherwise sing outside the shower, for example, lustily exclaim, "Take me out to the ballgame." And all because a king or a president once tried to sneak out early, or slip out for a pee. Substantial consequences from utterly insignificant origins.

I raise this theme because I recently realized that the primary "old standard," the classic textbook illustration of our preferences for Darwinian evolution, arose in the same manner—as an entrenched and ubiquitous example based on an assumed weight of historical tradition that simply does not exist. Several years ago, I made a survey of all major high school textbooks in biology. Every single one—no exceptions—began its chapter on evolution by first discussing Lamarck's theory of the inheritance of acquired characters, and then presenting Darwin's theory of natural selection as a preferable alternative. All texts then use the same example to illustrate Darwinian superiority—the neck of the giraffe.

The Tallest Tale

Giraffes, we are told, got long necks in order to browse the leaves at the tops of acacia trees, thereby winning access to a steady source of food available to no other mammal. Lamarck, the texts continue, explained the evolution of long necks by arguing that giraffes stretched and stretched during life, elongated their necks in the process, and then passed these benefits along to their offspring by altered heredity.

This lovely idea may embody the cardinal virtue of effort rewarded, but heredity, alas, does not operate in such a manner. A neck stretched during life cannot alter the genes that influence neck length—and offspring cannot reap any genetic reward from parental striving. We therefore prefer the Darwinian alternative, consistent with the Mendelian nature of heredity, that giraffes with fortuitously longer necks (in a varying population with a large range of neck lengths among individuals) will tend to leave more surviving offspring. These progeny will inherit the genetic propensity of their parents for greater height. This slow process, continued for countless generations, can lead to steady increase in neck length, so long as local environments continue to favor animals with greater reach for those succulent topmost leaves.

We often symbolize movements and beliefs by icons of clear meaning based on shared cultural histories. Thus, Americans may proclaim political affiliations by sporting a pin with a donkey or an elephant. More specifically, and among donkeys, a button with nothing but a saxophone identified "Friends of Bill" in Mr. Clinton's presidential campaign, just as a pin of a shoe with a hole (recalling a famous photo of a tired candidate) once rallied the supporters of Adlai Stevenson. Similarly, the tallest of mammals, sticking his neck up, stands for evolution, and particularly for Darwin's mechanism of natural selection. When Francis Hitching wrote a recent iconoclastic book, for evolution but against Darwinism, he chose as his title *The Neck of the Giraffe*—even though his text barely mentions the creature.

A story so often repeated should rise from firm foundations and bear both strong and graceful support throughout the length of construction. In short, this most familiar of all examples should, like the subject's own head, stand tall above everything else, buttressed by a device as supple and as well designed as "the neck of the giraffe." Or, to recall the second image of my opening sentence, and to quote from the greatest of all love poems (called, not inappropriately, the Song of Songs): "Thy neck is as a tower of ivory . . . This thy stature is like to a palm tree."

If, instead, we traced this ubiquitous example back to scraps of speculation, and discovered either no foundation at all, or a funny little point of origin equivalent to a king in need of a bathroom break, then we might learn two lessons of potential import: first, that repetition need not correlate with truth value, and that even the most pious certainties should be periodically scrutinized right down to their foundations; and second, that the current importance and utility of a phenomenon gives us no particular insight into the circumstances of its historical origin.

When we look to presumed sources of origin for competing evolutionary explanations of the giraffe's long neck, we find either nothing at all, or only the shortest of speculative conjectures. Length, of course, need not correlate with importance. Garrulous old Polonius, in a rare moment of clarity, reminded us that "brevity is the soul of wit" (and then immediately vitiated his wise observation with a flood of woolly words about Hamlet's madness). Many of the most famous Bible stories occupy only a verse or two, while lists of laws and begats go on for pages.

Yet length must bear at least a rough relationship to perceived depth of meaning. Few authors will write chapters on matters deemed trivial and then devote only a line to their own most treasured theme—if only because readers will then be unable to weigh the relative importances properly. I feel quite confident that the authors of the Old Testament did grant greater meaning to their genealogies and laws—the basis of order and power in their own society, after all—than to the story of Jonah and the whale (the shortest chapter in one of the Bible's shortest books). Our contemporary inversion—for fish stories now trump long lists of begats for unknown people with unpronounceable names—merely illustrates my chief point that current utility must be separated from historical origin in any judgment of importance or meaning.

The giraffe's neck just wasn't a big issue for the founders of evolutionary theory—not as a case study for arguing about alternative mechanisms, not for anything much at all. No data from giraffes then existed to support any particular theory of causes over another, and none exist now. Absence of data rarely stops an imaginative scientist from speculating, I admit, but you can generate just so many words before a paucity of information dries up your thoughts. And no decent natural historian—let us hope, at least—will use a purely speculative case as a primary illustration of a central theory.

Lamarck did mention giraffe necks as a putative illustration of evolutionary enlargement by the inherited effects of lifetime effort. But his entire discussion runs for one paragraph in a chapter filled with much longer examples that Lamarck obviously regarded as far more important. Lamarck had this to say—and absolutely nothing more—about giraffe necks, a few lines of speculation never intended as the centerpiece of a theory:

> It is interesting to observe the result of habit in the peculiar shape and size of the giraffe: this animal, the tallest of the mammals, is known to live in the interior of Africa in places where the soil is nearly always arid and barren, so that it is obliged to browse on the leaves of trees and to make constant efforts to reach them. From this habit, long maintained in all the individuals of the race, it has resulted that the animal's fore-legs have become longer than its hind-legs, and that its neck is lengthened to such a degree that the giraffe, without standing up on its hind-legs, can raise its head to a height of six meters [from Lamarck's classic 1809 work, *Philosophie zoologique*, volume 1, page 122, my translation].

This paragraph contains a giveaway statement—but you have to know the eighteenth-century literature to spot the clue—proving that Lamarck cared little about giraffes, and therefore didn't grant this throwaway example much weight. Public zoos in the modern sense did not then exist in Europe, and few private menageries (usually maintained for royal patrons) had ever housed giraffes. Some travelers had seen giraffes in the wild, and many visitors had viewed them on display in Cairo. Giraffes had been known to Europeans since classical times, when Roman emperors included them in public slaughters at the Coliseum. Some reports had claimed, as Lamarck affirms, that the giraffe's front legs greatly exceeded the back legs in height. In fact, both pairs of legs are equally tall. The impression of greater frontal height arises from the pronounced rearward slope of the giraffe's back, a consequence of the massive muscles and spinal projections needed up front to support the huge neck. The most reliable sources available in Lamarck's day had adequately established the equal length of fore and rear legs, and had dismissed the old myth of superior frontal elongation. Thus, if Lamarck repeated the old legend in his single paragraph about giraffes, he couldn't have read the literature thoroughly.

The example gained no particular steam as English writers explained Lamarck's theory to their countrymen. Lyell's remarkably fair exposition in opposition—given in the second volume of his *Principles of Geology* in 1832, and the source of most early English contact with Lamarck's theory—quoted the example in abridgment, and made no further comment. In his famous series of lectures to workingmen *(On Our Knowledge of the Causes of the Phenomena of Organic Nature),* published in 1863 as the first great popular exposition of Darwinism, T. H. Huxley omitted giraffes entirely, and illustrated Lamarck's theory with two examples emphasized by the Frenchman himself: the blacksmith's strong right arm, putatively inherited by his sons; and the long legs and webbed feet of shorebirds, presumably evolved to avoid submersion or slipping in muddy ponds or flowing waters.

When we turn to the horse's mouth, the first edition of Darwin's *Origin of Species* (1859), we find no mention whatever of the giraffe's neck as an illustration of natural selection. Interestingly—and proving my point with panache—Darwin does cite the giraffe in just the context usually assumed for the legend of the neck: for a speculative story about the efficacy of nat-

Buffon's illustration of a giraffe, showing that Lamarck's contemporaries clearly knew and depicted the equal heights of fore and hind legs.

ural selection. But, in this passage, Darwin treats the giraffe's opposite end, and tells a tale about the tail. Moreover—because Darwin did not much favor the fatuous "just-so story" mode for illustrating natural selection by plausible speculation alone—his story about the giraffe's tail occupies only a passing paragraph.

Darwin raises this tale to contend that natural selection has sufficient power to explain "organs of trifling importance." The giraffe's tail, he argues, works primarily as a flyswatter. One might regard such a function as too trivial to fall under the purview of a mechanism based on differential survival (can swatting flies really become a matter of life or death?). Darwin replies:

> The tail of the giraffe looks like an artificially constructed fly-flapper; and it seems at first incredible that this could have been adapted for its present purpose by successive slight modifications, each better and better, for so trifling an object as driving away flies; yet we should pause before being too positive even in this case, for we know that the distribution and existence of cattle and other animals in South America absolutely depend on their power of resisting the attacks of insects: so that individuals which could by any means defend themselves from these small enemies, would be able to range into new pastures and thus gain a great advantage. It is not that the larger quadrupeds are actually destroyed (except in some rare cases) by the flies, but they are incessantly harassed and their strength reduced, so that they are more subject to disease, or not so well enabled in a coming dearth to search for food, or to escape from beasts of prey.

Darwin does mention the giraffe's neck in a single line, but (ironically) for a purpose *opposite* to illustrating the power of natural selection in shaping organisms for particular utilities. In a closing section on evidence for evolution provided by homology, or retention of common ancestral structures in descendants of markedly divergent functional design, Darwin notes that the giraffe builds its remarkable neck not by adding new vertebrae, but by elongating the same seven bones present in the necks of virtually all mammals. Thus, history limits the power of natural selection by constraining adaptive solutions to the confines of inherited designs. He writes:

> The framework of bones being the same in the hand of a man, wing of a bat, fin of the porpoise, and leg of the horse,—the same number of vertebrae forming the neck of the giraffe and of the elephant,—and innumerable other such facts, at once explain themselves on the theory of descent with slow and slight successive modifications.

In his subsequent, and longest, book, the two-volume *Variation of Animals and Plants Under Domestication* (1868), Darwin does finally introduce the giraffe's neck in a discussion of natural selection. But again, and ironically given the later codification of the case as our canonical just-so story in the speculative tradition, Darwin does not cite "the neck of the giraffe" to tell a fatuous story about presumed adaptive advantages. Rather, he raises the example to discuss a more subtle issue central to the validity of natural selection as a general explanation of evolution.

Even if we assume that the giraffe's neck evolved as an adaptation for eating high leaves, how could natural selection build such a structure by gradual increments? After all, the long neck must be associated with modifications in nearly every part of the body—long legs to accentuate the effect, and a variety of supporting structures (bones, muscles, and ligaments) to hold up the neck. How could natural selection simultaneously alter necks, legs, joints, muscles, and blood flows (think of the pressure needed to pump blood up from the heart to the giraffe's faraway brain)? In response to this problem, some critics had proposed that all relevant parts must be changed together in one fell swoop. Such suddenly coordinated modification would invalidate natural selection as a creative force because the desired adaptation would then arise all at once as a fortuitous consequence of internally gener-ated variation. (Moreover, Darwin adds, we have no evidence for a deus ex machina of such complexly and luckily coordinated variation—and the whole proposal smacks of desperation and special pleading.)

Darwin provides a cogent and subtle explanation (perhaps not thor-oughly satisfactory by current views, but entirely logical and coherent). Interestingly, his proposal embodies the theme of this essay—the need to dissociate current utility from historical origin. A giraffe's current function-ing may require coordinated action of all parts that support the long neck, but these features need not have evolved in lockstep. If the neck grows by ten feet all at once, then every supporting bit of anatomy must be in place. But if the neck elongates an inch at a time, then the full panoply of support-ing structures need not arise at every step. The coordinated adaptation can be built piecemeal. Some animals may slightly elongate the neck, others the legs; still others may develop stronger neck muscles. By sexual reproduction, the favorable features of different organisms may be combined in offspring.

In developing this general explanation by using the giraffe as a putative example, Darwin does engage in conjectural biology. But I would defend this mode of speculation as a device utterly different from telling fatuous stories. When scientists need to explain difficult points of theory, illustration by hypothetical example—rather than by total abstraction—works well (perhaps indispensably) as a rhetorical device. Such cases do not function as "speculations" in the pejorative sense—as silly stories that provide no insight into complex mechanisms—but rather as idealized illustrations to exemplify a difficult point of theory. (Other fields, like philosophy and the law, use such conjectural cases as a standard device.)

In thus invoking the giraffe as a proper exemplification, Darwin does embed a line within his text about adaptive advantages of reaching high. Taken out of context, this comment could be read as a premonition of silly speculations to come. But its role as part of a conjectural case to illustrate a more subtle point of theory should be clear in the following totality (from Darwin's 1868 book, volume 2, pages 220–221):

The Tallest Tale

With animals such as the giraffe, of which the whole structure is admirably coordinated for certain purposes, it has been supposed that all the parts must have been simultaneously modified; and it has been argued that, on the principle of natural selection, this is scarcely possible. But in thus arguing, it has been tacitly assumed that the variations must have been abrupt and great. No doubt, if the neck of a ruminant were suddenly to become greatly elongated, the fore-limbs and back would have to be simultaneously strengthened and modified; but it cannot be denied that an animal might have its neck, or head, or tongue, or fore-limbs elongated a very little without any corresponding modification in other parts of the body; and animals thus slightly modified would, during a dearth, have a slight advantage, and be enabled to browse on higher twigs, and thus survive. A few mouthfuls more or less every day would make all the difference between life and death. By the repetition of the same process, and by the occasional intercrossing of the survivors, there would be some progress, slow and fluctuating though it would be, towards the admirably coordinated structure of the giraffe.

I suspect that the giraffe's neck first became an explicit and contested issue within evolutionary theory when St. George Mivart, a fascinating rebel in many ways—as a devout Catholic in this Anglican land, and as an evolutionist firmly opposed to the mechanism of natural selection—published his 1871 critique of Darwinism, *The Genesis of Species*. Mivart did focus on the giraffe's neck, and he did present Darwin's supposed case in the form that has become canonical in modern high school textbooks—that is, as a speculative tale about natural selection. But note that Mivart wrote to *oppose* Darwinism, and that enemies tend to caricature and trivialize the doctrines they attack. Mivart stated:

At first sight it would seem as though a better example in support of "Natural Selection" could hardly have been chosen. Let the fact of the occurrence of occasional, severe droughts in the country which that animal has inhabited be granted. In that case, when the ground vegetation has been consumed, and the trees alone remain, it is plain that at such times only those individuals (of what we assume to be the nascent giraffe species) which were able to reach high up would be preserved, and would become the parents of the following generation.

For the sixth and last edition of the *Origin of Species* (1872), Darwin added the only chapter ever appended to his book—primarily to refute Mivart's attack. This new chapter does discuss giraffes extensively (though only to rebut Mivart), and may be a primary source for the legend as later developed (for almost all reprintings and subsequent versions, up to our own day, feature this sixth edition, and not the first edition of 1859, which did not mention giraffe necks in the context of natural selection at all).

When we read Darwin's careful words, however, we encounter yet another irony in our expanding list. The giraffe's neck supposedly supplies a crucial example for preferring natural selection over Lamarckism as a cause of evolution. But Darwin himself (however wrongly by later judgment) did not deny the Lamarckian principle of inheritance for characters acquired by use or lost by disuse. He regarded the Lamarckian mechanism as weak, infrequent, and entirely subsidiary to natural selection, but he accepted the validity of evolution by use and disuse. Darwin does speculate about the adaptive advantage of giraffes' necks, but he cites *both* natural selection *and* Lamarckism as probable causes of elongation. Thus, obviously, Darwin never regarded giraffe necks as an illustration for the superiority of natural selection over other valid mechanisms. He writes in two passages of the 1872 edition, marrying Lamarck with natural selection:

> By this process [natural selection] long continued . . . combined no doubt in a most important manner with the inherited effects of the increased use of parts, it seems to me almost certain that an ordinary hoofed quadruped might be converted into a giraffe.

The Tallest Tale

> In every district some one kind of animal will almost certainly be able to browse higher than the others; and it is almost equally certain that this one kind alone could have its neck elongated for this purpose, through natural selection and the effects of increased use.

We may summarize the main line of this complexly meandering tale as a list of ironies—invoking the technical definition of irony as a statement where, for humorous or sarcastic effect, the intended meaning of a word becomes directly opposite to the usual sense—as in "that's very smart!" for a proposal you regard as consummately dumb. In this story, none of the five historical facts arose by ironic intent. The irony occurs retrospectively, for each fact subverts the legend that "everyone knows" about tall giraffes—namely, that long necks for high leaves provide a splendid illustration for the superiority of Darwinian natural selection over Lamarckian use and disuse. The joke, in other words, is on the silly canonical legend as recounted in all modern textbooks.

1. Lamarck mentions giraffe necks in one passing paragraph of speculation within a chapter devoted to much longer examples regarded as far more important.
2. Darwin does not cite the case at all in the first edition of the *Origin of Species*. He does tell a giraffe story in the "just-so" mode, but from the opposite end—the tail rather than the neck. Darwin's only quick phrase about giraffe necks illustrates the contrary theme of inherited stability (retained number of neck vertebrae) rather than novel adaptation.
3. When Darwin, in his longer and more technical book of 1868, does discuss giraffe necks in the context of natural selection, he does not present the standard "just-so story" of pure speculation, but rather uses giraffes to exemplify the difficult and crucial issue of how gradualistic natural selection can build a complex adaptation of many coordinated parts (the neck *and* all the supporting structures).
4. Mivart, in attempting to refute Darwinism, tells the "just-so story" that would become traditional, but only to caricature a theory he opposes.

5. When Darwin responds to Mivart in the last edition of the *Origin of Species,* he does interpret giraffe necks as adaptations for feeding on high leaves, but he argues that natural selection worked in concert with Lamarckian forces! (So much for a "classic" illustration of why the giraffe's neck leads us to prefer Darwin over Lamarck.)

I don't know (but would love to find out) how and where the legend's modern form originated in such striking contrast to alleged historical sources. Henry Fairfield Osborn, the dominant paleontologist of his era, and longtime director of the American Museum of Natural History, gave the "standard" version in his popular book of 1918, *The Origin and Evolution of Life:*

> The cause of different bodily proportions, such as the very long neck of the tree-top browsing giraffe, is one of the classic problems of adaptation. In the early part of the nineteenth century Lamarck attributed the lengthening of the neck to the inheritance of bodily modifications caused by the neck-stretching habit. Darwin attributed the lengthening of the neck to the constant selection of individuals and races which were born with the longest necks. Darwin was probably right.

This version has held ever since. Readers may well ask why we should devote energy to tracing such historical arcana. Why not let sleeping dogs lie and silly legends propagate, especially if tall tales do no harm? I gave some theoretical reasons for interest earlier in this essay, but I also wish to stress a practical concern. If we choose a weak and foolish speculation as a primary textbook illustration (falsely assuming that the tale possesses a weight of history and a sanction in evidence), then we are in for trouble—as critics properly nail the particular weakness, and then assume that the whole theory must be in danger if supporters choose such a fatuous case as a primary illustration. For example, in his anti-Darwinian book cited earlier (and eponymously titled *The Neck of the Giraffe*), Francis Hitching tells the story in the usual form:

The Tallest Tale

The evolution of the giraffe, the tallest living animal, is often taken as classic evidence that Darwin was right and Lamarck wrong. The giraffe evolved its long neck, it is said, because natural selection choose those animals best able to feed off the highest treetops, where there is most food and least competition.

Hitching then adds: "The need to survive by reaching ever higher for food is, like so many Darwinian explanations of its kind, little more than a *post hoc* speculation." Hitching is quite correct, but he rebuts a fairy story that Darwin was far too smart to tell —even though the tale later entered our high school texts as a "classic case" nonetheless. Eternal vigilance, as they say, is the price of freedom. Add intellectual integrity to the cost basis.

As a closing point, we might excuse this thoughtless repetition of an old legend without presumed historical sanction, if later research had established the truth of the tale nonetheless. But when we turn to giraffes themselves, we encounter the final irony of this long story. Giraffes provide no established evidence whatsoever for how their undeniably useful necks evolved.

All giraffes belong to a single species, quite separate from any other ruminant mammal, and closely related only to the okapi (a rare, short-necked, forest-dwelling species of central Africa). Giraffes have a sparse fossil record in Europe and Asia, but ancestral species are relatively short-necked, and the spotty evidence provides no insight into how the long-necked modern species arose. (*The Giraffe: Its Biology, Behavior and Ecology,* by A. I. Dagg and J. B. Foster, gives an excellent and thorough account of all major aspects of giraffe biology.)

When we study the function of long necks in modern giraffes, we encounter an *embarras de richesses.* Almost anything important in the life of a giraffe involves some use of the remarkable neck. Giraffes surely employ their long necks (and their long legs, long faces, and long tongues) to reach high-growing acacia leaves. Giraffes thereby browse several feet of vegetation exploitable by no other ground-dwelling mammal. The champion giraffe reached an astonishing nineteen feet, three inches in height. Groves of African acacia trees (I have seen this phenomenon in the field) are often denuded below a sharp line representing the highest reach of local giraffes.

But giraffes also use their necks for other prominent and crucial activities. Male giraffes, for example, establish dominance hierarchies by frequent and prolonged bouts of "necking," or swinging their large neck into the body of an opponent. These contests are more than merely symbolic, as the long neck propels the head with substantial force, and the bony horns atop the head can inflict considerable damage upon contact. Dagg and Foster describe a bout between two males named Star and Cream:

> The two bulls . . . stood side to side, head to tail, close together, each with his legs apart under him for balance. Suddenly Star lowered his head and whipped it, horns foremost, at Cream's trunk, connecting with an impact that was heard easily from forty meters away. Cream lurched sideways, collected himself and returned the blow with his head, striking Star on the neck. Star then aimed at Cream's front legs and knocked them out from under him with a blow of his head.

Dagg and Foster then describe the serious finality of potential outcomes:

> The losing giraffe in such a struggle does not always escape so easily. His head may be gashed during a fight or he may be knocked to the ground unconscious . . . In such a contest in the Kruger National Park one of the contestants was killed. He had a large hole immediately behind one ear where his top neck vertebra had been splintered by a blow; part of the splinter had pierced the spinal cord.

Interestingly, giraffes fight predators (primarily lions) by kicking, but their sexual combats proceed by necking, never by kicking. Thus, this function of the neck may represent a specifically evolved behavior for a particular circumstance.

Giraffes also use their necks in several other ways: as a "lookout tower" to spot predators and other dangers, and as a device to increase surface area and shed heat (giraffes, unlike other large African mammals, do not seek shade and can remain in the sun). Both these functions have been viewed by prominent scientists as a chief reason for the evolution of long necks. In addition, giraffes deftly shift their center of gravity by appropriate movements of the neck—and these maneuvers are crucial to a wide range of activities, including rising from a lying position, running, and climbing fences and other barriers.

We may now return to the central theme of this essay—the dissociation of current utility from historical origin—and understand why the giraffe's neck cannot provide a proof for any adaptive scenario, Darwinian or otherwise. Giraffes do use their long necks to browse leaves at the tops of acacia trees—but such current function, no matter how vital, does not prove that the neck originally evolved for this purpose. The neck may have first lengthened in the context of a different use, and then been coopted for better dining when giraffes moved into the open plains. Or the neck may have evolved to perform several functions at once. We cannot learn the reasons for historical origin simply by listing current uses.

When we consider the full range of current function, we can be fairly confident that some uses must be secondary, and cannot therefore be the source of historical origin. I can't imagine, for example, that long necks evolved to help giraffes maneuver in running, jumping, and getting up—because the problem only arose when giraffes acquired a long neck in the first place, and solutions to problems can't be causes of the problem.

But other functions may well be original—and the famous reaching for leaves could arise as a largely secondary effect. Since natural selection works fundamentally by differential reproductive success, and since sexual combat so often acts as a primary determinant of this basic Darwinian benefit, we could state a plausible case for regarding sexual success as the chief adaptive reason for evolving long necks, with the much-vaunted browsing of leaves as a distinctly secondary consequence. In short, we have no basis for any firm assertion about the most famous inquiry among Darwinian just-so stories: How did the giraffe get its long neck?

This essay therefore features a double whammy in pursuit of a primary theme—the dissociation of current utility from historical origin. In the realm of ideas, current invocation of the giraffe's neck as the classic case of Darwinian evolution does not grow from firm and continuous historical roots. The standard story, in fact, is both fatuous and unsupported. In the realm of giraffes, current use of maximal mammalian height for browsing acacia leaves does not prove that the neck evolved for such a function. Several reasonable alternative scenarios exist, and we have no evidence for preferring any plausible version over another. *Caveat lector.*

Why, then, have we been bamboozled into accepting the usual tale without questioning? I suspect two primary reasons: we love a sensible and satisfying story, and we are disinclined to challenge apparent authority (like textbooks!). But do remember that most satisfying tales are false. The seventh-inning stretch predated Mr. Taft, and the story of kingly rising before the Hallelujah Chorus has no established foundation either. Polonius may have been an old bore, but he did give Laertes some good advice in the famous speech that Laertes surely failed to process because he was trying so hard to leave town. Among other tidbits, Polonius emphasized the importance of overt appearance—and we would do well to remember his counsel. Darwinian evolution may be the most truthful and powerful idea ever generated by Western science, but if we continue to illustrate our conviction with an indefensible, unsupported, entirely speculative, and basically rather silly story, then we are clothing a thing of beauty in rags—and we should be ashamed, "for the apparel oft proclaims the man."

17

B R O T H E R H O O D   B Y

I N V E R S I O N

(O R ,   A S   T H E   W O R M   T U R N S)

AS HAMLET, IN THE MOST CELEBRATED SOLILOQUY OF ENGLISH LITERA-
ture, weighs the relative values of life and death, he describes the attraction
of suicide ("not to be") as an escape from *active* insults, including "the
oppressor's wrong, the proud man's contumely." But writers and intellectuals worry far more about an opposite fate on life's potential "sea of trou-
bles"—erasure and oblivion, the pain of being simply ignored. Samuel
Johnson, as recorded by Boswell, expressed this silent arrow of outrageous
fortune in a famous aphorism: "I would rather be attacked than unnoticed.
For the worst thing you can do to an author is to be silent as to his works."

I therefore felt special poignancy when I recently read an anecdote
about the last years of a great English physiologist, Walter H. Gaskell
(1847–1914). After a distinguished career of solid experimental work on the
function of the heart and nervous system, Gaskell switched gears and
devoted the entire second half of his professional life (from 1888 on) to pro-

moting and defending an idiosyncratic theory for the origin of vertebrates. The last paragraph of Gerald L. Geison's long article on Gaskell in the *Dictionary of Scientific Biography* reads:

> His final years were clouded . . . by a feeling that his deeply loved theory of the origin of vertebrates was not receiving a fair hearing. Even at Cambridge, where Gaskell lectured on the topic until his death, his audience decreased over the years until, near the end, the poignant scene is drawn of Gaskell closing his course by shaking hands with a lone remaining auditor.

We may grieve for Gaskell's personal fate as an intellectual pariah; but, truth to tell, he had been pushing a pretty nutty theory for the origin of vertebrates. Gaskell believed with all his soul, and with a striking absence of critical questioning, that the evolution of animal life must follow a single pathway of progressive advance mediated by an increasing elaboration of the brain and nervous system. Gaskell wrote in his major work of 1908, *The Origin of Vertebrates* (the source of all quotes from Gaskell in this essay):

> We can trace without a break, always following out the same law, the evolution of man from the mammal, the mammal from the reptile, the reptile from the amphibian, the amphibian from the fish, the fish from the arthropod [insects and their allies], the arthropod from the annelid [segmented worms], and we may be hopeful that the same law will enable us to arrange in orderly sequence all the groups in the animal kingdom.

Gaskell identified this controlling principle of linear advance as the "law of the paramount importance of the development of the central nervous system for all upward progress." In a rhetorical flourish, he then inverted the Preacher's famous argument (Ecclesiastes 9:11) for randomness and aimless change without direction: "The law of progress is this—The race is not to the swift, nor to the strong, but to the wise."

Advocates for a single line of progress encounter their greatest stumbling block when they try to find a smooth link between the apparently disparate designs of invertebrates and vertebrates. In addressing this old problem, Gaskell adopted the standard strategy of linear progress theorists from time immemorial: identify the most complex invertebrate and attempt to forge a link with the simplest vertebrate. Gaskell, again following tradition, selected arthropods as the invertebrate pillar for his bridge, and then tried to build the span by his law of neurological complexification. He wrote:

> This consideration points directly to the origin of vertebrates from the most highly organized invertebrate group—the Arthropoda—for among all the groups of animals living on the earth in the present day they alone possess a central nervous system closely comparable in design with that of vertebrates.

So far, so conventional. Gaskell's theory becomes idiosyncratic, and a bit bizarre, in his chosen mode for forging the improbable link of arthropod to vertebrate. Among the plethora of prominent differences between these phyla, one central contrast has always served as a focus for discussion, and a chief impediment to any linear scheme. Arthropods and vertebrates share some broad features of general organization—elongated, bilaterally symmetrical bodies, with sensory organs up front, excretory structures in the back, and some form of segmentation along the major axis. But the geometry of major internal organs could hardly be more different, thus posing the classical problem that has motivated several hundred years of dispute and despair among zoologists.

Arthropods concentrate their nervous system on their ventral (belly) side as two major cords running along the bottom surface of the animal. The mouth also opens on the ventral side, with the esophagus passing between the two nerve cords, and the stomach and remainder of the digestive tube running along the body *above* the nerve cords. In vertebrates, and with maximal contrast, the central nervous system runs along the dorsal (top) surface as a single tube culminating in a bulbous brain at the front end. The entire digestive system then runs along the body axis *below* the nerve cord. (The accompanying figure from Gaskell's book illustrates this cardinal difference in an unconsciously amusing way.) But could evolution (or a sensible divine creator, for that matter) turn an arthropod with belly above nerve cords into a superior vertebrate with brains on top and gut below?

Gaskell proposed a pretty wild scheme for such a transformation, and his loss of respect (and students) followed his inability to construct a cogent defense. Gaskell argued that the dorsal gut of arthropods evolved into the vertebrate brain and spinal cord as a proliferation of nervous tissue fueled the upward march of progress. This new nervous tissue began to surround the old gut, eventually choking off all digestive function like a strangler fig around a host tree, or an anaconda squeezing the lifeblood from a pig. The modern vertebrate brain surrounds the old arthropod stomach, thus explaining the ventricles—the interior spaces between the folds of the

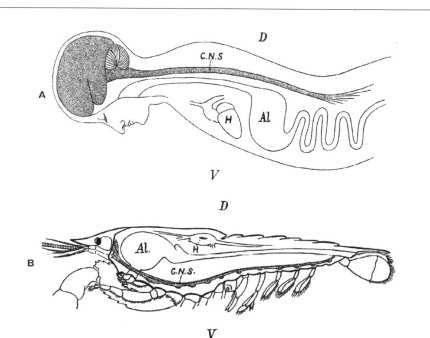

Gaskell's unintentionally humorous drawing of the basic anatomy of insects and vertebrates. Note how the central nervous system (labeled CNS) lies above the gut (labeled AL) in vertebrates, but below the gut in arthropods.

brain—as remnants of an ancestral digestive space. Similarly, the central canal of the spinal cord represents the old arthropod intestine, now surrounded by nervous tissue.

But this putative solution only engendered an even more troubling problem: If the arthropod gut became the vertebrate nervous system, then what organ can serve as a precursor for the vertebrate gut? This problem stymied Gaskell, and he opted for a deus ex machina that eventually satisfied no one but himself (and perhaps his one last auditor): the vertebrate digestive tube simply arose *de novo*, to suit an obvious need. Gaskell concluded:

> Vertebrates arose from ancient forms of arthropods by the formation of a new alimentary canal, and the enclosure of the old canal by the growing central nervous system.

Can we extract any message from Gaskell's failed theory beyond a stodgy, if appropriate, warning about the virtues of caution and sobriety? I certainly think so, for I have long held, and expressed as a mainstay of these essays, that when fine scientists devote their careers to theories later judged nutty or crazy, interesting and instructing reasons always underlie the paradoxical advocacy. This principle certainly applies in Gaskell's case because we can identify both a generally constraining bias and a personally compelling reason that drove Gaskell to the odd idea of stomachs turning into brains and new guts arising from nothing but inchoate potentiality.

Gaskell's dubious but unquestioned conviction about linear progress served as the general bias that led him to propose an almost alchemical scheme of transmutation from arthropod to vertebrate. But an understanding of the history of this subject also reveals a particular reason that interacted with his general conviction about progress to lead him down a path of increasing irrelevance and loneliness. In short, Gaskell proposed his own nutty theory because he couldn't abide the older and standard account, also judged by history as a prime case of nuttiness, for linking arthropods and vertebrates.

Think about the basic contrast, and the most obvious way to produce alignment. Arthropods grow ventral nerve cords with the gut above; vertebrates develop with a dorsal nerve tube and a gut below. *Presto turno*—and one becomes the other. Why not just invert a segmented worm or an insect to produce the vertebrate pattern? Turn a bug on its back (as Kafka did, come to think of it, when he changed his protagonist into a roach in "The Metamorphosis"), and the internal geometry of a vertebrate emerges— nerves above guts.

I don't mean to be frivolous or cavalier about complex and serious matters. All participants in the history of this debate know perfectly well that an inverted worm or insect doesn't become a vertebrate, *tout simple* and all nice and clean. More than a few knotty problems and inconsistencies remain. To cite the dilemma most widely discussed in the literature, the esophagus of an inverted bug runs upward through the nervous system (right in the area that will become the vertebrate brain), to emerge at a mouth on top of the head. Clearly this will not do (and has not done in any real vertebrate)! So the

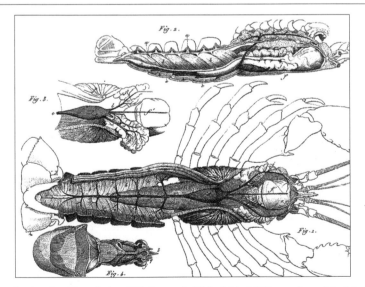

French zoologist Geoffroy Saint-Hilaire embellished his 1822 treatise with this drawing of a lobster's anatomy. In the top figure, the arthropod is depicted on its back to illustrate how closely its structure resembles that of an inverted vertebrate.

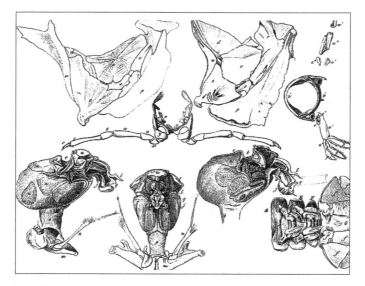

This plate from Geoffroy's treatise includes, in the center of the plate, a lobster segment with a pair of legs extended, meant to emphasize its supposed structural affinity with a vertebra with attached ribs.

inversion theory for deriving vertebrates from arthropods must argue that the old brain-piercing mouth atrophied and closed up, while a new ventral mouth developed at the front end of the vertebrate gut. Forming a new mouth at the end of an old tube may not be quite so bold or improbable as building an entirely new gut from nothing (as Gaskell's theory required), but no evidence for such a scenario exists either, and the whole tale smacks of fatuous special pleading to save an otherwise intriguing idea.

In any case, I am not spinning an abstract fairy tale as a hypothetical alternative to Gaskell's solution. The inversion theory has a long and fascinating history in the discussion of vertebrate origins. The founding version dates to the early nineteenth century and became the centerpiece of a movement often called "transcendental biology," and centered on the attempt to reduce organic diversity to one or a very few archetypal building blocks that could then generate all actual anatomies as products of rational laws of transformation. Some of Europe's greatest thinkers participated in this grand, if flawed, enterprise. Goethe, Germany's preeminent poet-scientist, tried to explain the varied parts of plants as different manifestations of an archetypal leaf. In France, Étienne Geoffroy Saint-Hilaire attempted to portray the skeleton of vertebrates as a set of modifications upon an archetypal vertebra.

In the 1820s, Geoffroy extended his ambitious program to include annelids and arthropods under the same rubric. With boldness verging on a mania too sweeping to be entirely right but also too ingenious to be completely wrong, he argued that arthropods also built their bodies on a vertebral plan, but with one central difference. Vertebrates support their soft parts with an internal skeleton, but insects, with their external skeletons, must live *within* their own vertebrae (a reality, not a metaphor, for Geoffroy). This comparison led to other strange consequences, all explicitly defended by Geoffroy, including the claim that a vertebrate rib must represent the same organ as an arthropod leg—and that insects must therefore walk on their own ribs!

Geoffroy also recognized that the opposite orientations of gut and nervous system posed a problem for his claim that insects and vertebrates represent different versions of the same archetypal animal—and he proposed the first account of the inversion theory to resolve this threat to unification. Geoffroy's initial version of 1822 makes much more sense than the later evolutionary scenarios of linear transformation that so enraged Gaskell. Geoffroy was an early evolutionist in these decades before Darwin, but he did not devise the inversion theory as a genealogical proposition—that is, he did not argue that an arthropod ancestor evolved directly into a primitive vertebrate by turning over. Geoffroy pursued the quite different aim of establishing a "unity of type" that could generate both arthropods and vertebrates from the same basic blueprint.

He then argued, quite cogently within his own framework, that this grand Platonic blueprint paid scant attention to such "insignificant" questions of nitty-gritty daily reality as which side of a universal design happened to point toward the sun. The single grand design includes a gut in the middle and the main nerve cords somewhere on the periphery. Arthropods orient this peripheral region down and away from the sun—so we call their nerve cords ventral. But vertebrates orient their spinal cord up and toward the sun—so we call the same structure dorsal in our own kin. In other words, arthropods and verterbates express one common design in two orientations, insignificantly inverted with respect to an external axis of sunlight and gravity.

But later evolutionary theorists of linear progress had to advance the overtly physical and historical claim that an ancestral lineage of arthropods actually turned over to become the first vertebrates (for the classic statement of the inversion theory in this genealogical form, see William Patten, *The Grand Strategy of Evolution,* 1920). Gaskell could not abide this indecorous version of his beloved linear progress theory. He could not bear to imagine that the grand procession from jellyfish to man, orchestrated by an ever-increasing mass of nervous tissue, once paused in its stately and orderly march toward human consciousness in order to execute a fancy little flip, a clever jig of inversion, just at the sublime and definitive moment of entrance into the vertebral home stretch.

Gaskell therefore had to keep his stately soldiers upright and uniformly oriented throughout their journey toward the human pinnacle—and he fulfilled this need by crafting the vertebrate brain and spinal cord from an arthropod digestive tube, while forming a completely new gut below. By this device, he could keep tops on top and bottoms at the bottom throughout the linear history of animal life, while placing nerves below the gut in arthropods, but above the gut in vertebrates. Gaskell thought that his move would rescue the theory of linear progress, with its necessary transition of arthropod into vertebrate, from the absurdities of the old inversion theory. "How is it then," he wrote, "that this theory has been discredited and lost ground? Simply, I imagine, because it was thought to necessitate the turning over of the animal." Gaskell therefore invented his peculiar alternative as a refutation of the venerable inversion theory. He wrote of the first vertebrate: "If the animal is regarded as not having been turned over . . . then the ventricles of the vertebrate brain represent the original stomach, and the central canal of the spinal cord the straight intestine of the arthropod ancestor."

How ironic. In order to avoid the "nutty" theory of inversion, Gaskell invented the even odder notion of stomachs turning into brains with new guts forming below. No wonder, then, that later biologists cast a plague on both speculative houses and opted instead for the obvious alternative: arthropods and vertebrates do not share the same anatomical plan at all, but rather represent two separate evolutionary developments of similar complexity from a much simpler common ancestor that grew neither a discrete gut nor a central nerve cord. After all, we now know that arthropods and vertebrates have been separated for more than 500 million years, and that "simpler" arthropods did not turn into "complex" vertebrates at some halfway point on a march to a single evolutionary apex.

Furthermore, this sensible idea of independent derivation meshed beautifully with the triumph, from the 1930s on, of a strict version of Darwinism based on the near ubiquity of adaptive design built by natural selection with little constraint imposed by strictures of a common anatomic groundplan like Goethe's leaf or Geoffroy's vertebra. If adaptation and natural selection wield such unimpeded power over the fate of each evolutionary sequence, why search for deeper commonalities in lineages long separate? Arthropods and vertebrates do share several features of functional design. But those similarities only reflect the power of natural selection to craft optimal structures independently in a world of limited biomechanical solutions to common functional problems—an evolutionary phenomenon called convergence.

After all, if you want to fly, you have to develop wings of some sort, because nothing else can work. Bats, birds, and pterosaurs (flying reptiles of dinosaur times) all evolved wings independently because natural selection knows no other solution, and holds the capacity to build such intricate convergences as independent illustrations of its predominant power. Therefore, if both arthropods and vertebrates evolved guts and nerves in reversed orientations, why worry about different expressions of a common constraint? The two phyla have been separate for half a billion years and undoubtedly evolved their digestive and neurological organs along separate pathways of adaptation.

This new consensus seemed so compelling that Ernst Mayr, the dean of modern Darwinians, opened the ashcan of history for a deposit of Geoffroy's ideas about anatomical unity. We now appreciate the immense power of natural selection to build and rebuild every feature; to change, and then to alter again, nearly every nucleotide of every gene in the interest of better adaptation. Lineages that have been separate for 500 million years cannot possibly retain enough genetic identity to encode any important common constraint of design. In his epochal book of 1963, *Animal Species and Evolution,* Mayr wrote:

> In the early days of Mendelism there was much search for homologous genes that would account for such similarities. Much that has been learned about gene physiology makes it evident that the search for homologous genes is quite futile except in very close relatives.

The verdict of history had descended. Gaskell had proposed a bizarre theory to reject Geoffroy's union of arthropods and vertebrates by inversion. But Geoffroy's theory turned out to be quite weird enough all by itself. Evolutionary studies would finally abandon such romantic nonsense and move into the light of unimpeded natural selection.

Except for one small matter. Darwin himself told us in his last book *(The Formation of Vegetable Mould Through the Action of Worms)* that we should never underestimate the collective power of worms on the move. Our general culture also recognizes two primary metaphors, one inorganic and one organic, for the reversal of received opinion. Well may traditionalists fear the turning of these two objects: tables and worms. The inversion of a humble worm, especially when disturbed, may bring down empires. Shakespeare told us that "the smallest worm will turn being trodden on." And Cervantes wrote in his author's preface to *Don Quixote* that "even a worm when trod upon, will turn again."

How wonderfully symbolic and real in the double meaning. Geoffroy proposed a theory to unite the architecture of complex animals by comparing vertebrates with segmented worms and arthropods turned over. This theory for the archetype of complex animals became, instead, the archetype of nutty ideas in biology. But turning worms also serve as a leading cultural metaphor for upheaval of accepted ways and thoughts. I have always loved the boldness of Geoffroy's theory, but I never dreamed that he might have been right—even though I have long embraced, as a centerpiece of my own career, his larger view about the importance of inherited architectural pathways as constraints upon the optimizing power of natural selection. Well, the worm turned twice during the past year or two—in both actual and symbolic styles. Geoffroy, it seems, was correct after all—not in every detail, of course, but at least in basic vision and theoretical meaning. And the triumph of this surprise, the inversion of nuttiness to apparent truth, stands as a premier example of the most exciting general development in evolutionary theory during our times.

I published my first technical book, *Ontogeny and Phylogeny,* in 1977. I took pride in this long work on the relationship between embryology and evolution, but also became quite frustrated because we then knew so preciously little about the potential key to a resolution: the genetic basis of development. How does the genetic code help to orchestrate this greatest miracle of everyday biology—the regular and usually unerring production of adult complexity from the apparent formlessness of a tiny fertilized egg? We knew practically nothing, but we assumed (as documented above) that the major animal phyla, all evolutionarily separate for at least 500 million years, could share no constraining common plan or genetic architecture. Pure Darwinism reigned triumphant, and natural selection had built each basic anatomy for its own adaptive utility.

But we can now determine, easily and relatively cheaply, the detailed chemical architecture of genes; and we can trace the products of these genes (enzymes and proteins) as they influence the course of embryology. In so doing, we have made the astounding discovery that all complex animal phyla—arthropods and vertebrates in particular—have retained, despite their half-billion years of evolutionary independence, an extensive set of common genetic blueprints for building bodies. Many similarities of basic design among animal phyla, once so confidently attributed to convergence, and viewed as testimony to the power of natural selection to craft exquisite adaptation, demand the opposite interpretation that Mayr labeled as inconceivable: the similar features are homologies, or products of the same genes, inherited from a common ancestor and never altered enough by subsequent evolution to erase their comparable structure and function. The similarities record the constraining power of conserved history, not the architectural skills of natural selection independently pursuing an optimal design in separate lineages. Vertebrates are, in a certain sense, true brothers (or homologs)—and not mere analogs—of worms and insects.

Examples of this primary reversal of standard theory have been accumulating for the past fifteen years. In the first pathbreaking case, the homeotic genes of insects, responsible for specifying the separate identities of segments along the main body axis (by orchestrating the growth of antennae, mouthparts, legs, and so on in their proper places), were also discovered, in minimally altered form, in vertebrates. (The homeotic genes were first recognized by oddball mutants with body parts in the wrong places—legs growing out of the head where antennae should be, for example. In *Drosophila,* the homeotic genes occur in two arrays on a single chromosome. Interestingly, in vertebrates, these same arrays exist in multiple copies, as four sequences on four separate chromosomes.) These vertebrate homologs do not control the basic segmentation of the vertebral column (so insect segments are not simple homologs of vertebrae, as Geoffroy had originally proposed). But the homeotic genes of vertebrae do regulate the embryonic segmentation of the mid- and hind brain, and they do strongly influence other important repetitive structures, including the positioning of cranial nerves along the body axis.

A second case then seriously compromised the classic textbook example of convergence—the paired eyes of three great phyla: vertebrates, arthropods (with the multiply faceted eye of flies as a primary example), and mollusks (particularly the complex lens eye of squid, so similar in function to our own, but built of different tissues). We had always assumed that eyes in the three phyla evolved three separate times, in complete independence, because they differ substantially in basic anatomy. And we viewed this supposed convergence as a premier example of natural selection's power to produce organs of similar and optimal function, but built from different materials and evolved from entirely separate starting points. But we now know that eyes in all three phyla share an inherited embryological pathway largely orchestrated by a gene (called *Pax-6* in its vertebrate form) retained in all these phyla from a common ancestor, and remaining similar enough to work interchangeably (for the fly version will induce eyes in vertebrates, and vice versa). The end results vary substantially (the multifaceted fly eye is not homologous with our single-lens eye), but the embryological blueprints share a common ancestry, and the eyes of different phyla can no longer be viewed as an example of pure convergence.

The reversal of opinion during the past decade has been astonishing. Mayr argued that we shouldn't even bother to look for genetic homology and shared embryological pathways between distinct phyla. We have now moved to the opposite pole of being surprised when we identify a basic gene of developmental architecture in *Drosophila* and then do *not* find a homolog in vertebrates. Charles B. Kimmel began a recent paper on this subject by writing: "We have come to find it more remarkable to learn that a homolog of our favorite regulatory gene in a mouse is not, in fact, present in *Drosophila* than if it is, given the large degree of evolutionary conservation in developmentally acting genes."

Still, I guess I haven't fully accommodated to the change—even though the new perspective suits my hopes and fuels my theoretical prejudices so well—for I never dreamed that my all-time favorite theory of interphyletic union, Geoffroy's hypothesis of inversion, could possibly be right as well. The basic structuring from front end to back? Fine. Eyes? Why not? But the arthropod belly as the vertebrate back? Kind of silly, however intriguing.

Except that Geoffroy's inversion theory, appropriately re-expressed in the language of modern genetics and developmental biology, turns out to be true. In several papers, published during the past two years, and based on

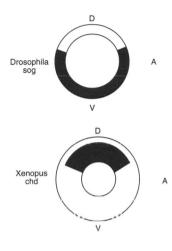

Drosophila
sog

Xenopus
chd

This figure from De Robertis and Sasai's article shows how the blackened ventral area (labeled V) forming the central nerve cords in the fruit fly Drosophila *corresponds with the dorsal area (also blackened and labeled D) forming the central nerve cords in the amphibian vertebrate* Xenopus.

work done primarily in the laboratories of Eddy M. De Robertis at UCLA and Ethan Bier at the University of California, San Diego, all essentials of Geoffroy's theory have been strikingly affirmed in contemporary terms (see especially Holley et al., 1995; De Robertis and Sasai, 1996; François et al., 1994; and François and Bier, 1995; all in the bibliography).

Geoffroy's vindication began with the sequencing of a vertebrate gene called *chordin.* In the toad *Xenopus* (but working, so far as we know, in a similar manner in all vertebrates), the *chordin* gene codes for a protein that patterns the dorsal (top) side of the developing embryo, and plays an important role in formation of the dorsal nerve cord. When these scientists searched for a corresponding gene in *Drosophila,* they discovered, to their surprise, that *chordin* shares sufficient similarity with *sog* to make a confident claim for common ancestry and genetic homology. But *sog* is expressed on the ventral (bottom) side of *Drosophila* larvae, where it acts to induce the formation of ventral nerve cords. Thus, the same gene by evolutionary ancestry builds both the dorsal nerve tube in vertebrates and the ventral nerve cords of *Drosophila*—in conformity with Geoffroy's old claim that vertebrate backs are arthropod bellies, and that the two phyla can be brought into structural correspondence by inversion.

This intriguing fact cannot affirm Geoffroy's inversion theory by itself, but De Robertis and colleagues then sealed the case with two additional discoveries. First, they found that a major gene responsible for specifying the dorsal side of flies (and called *decapentaplegic,* or *dpp*), has a vertebrate homolog (called *Bmp-4*) that patterns the ventral side of *Xenopus*—another reversal consistent with Geoffroy's hypothesis. Moreover, the entire system seems to work in the same way—but inverted—in the two phyla. That is, *dpp,* diffusing from the top to the bottom, can antagonize *sog* and suppress the formation of the ventral nerve cords in *Drosophila*—while *Bmp-4* (the homolog of *dpp*) diffusing from the bottom to the top, can antagonize *chordin* (the homolog of *sog*) and suppress the formation of the dorsal nerve cord in vertebrates. (The preceding figure, taken from the original publications, shows these relationships better than words can convey.)

Second, these scientists also found that the fly gene can work in humans, and vice versa. Vertebrate *chordin* can induce the formation of ventral nerve tissue in flies, while fly *sog* can induce dorsal nerve tissue in vertebrates. I regard these three discoveries as forming a tight and well-documented case for Geoffroy's old theory of inversion.

Moreover, current results vindicate Geoffroy's version, not the later scenarios of linear evolution. These data do not support the silly notion that, at a defining moment in the march of evolutionary progress, an arthropod literally flipped over to become the first vertebrate. Rather, as Geoffroy argued so long ago, the two phyla share a common architecture, but in reversed arrangement. In evolving separately from common ancestry, vertebrates oriented the shared design in one manner, annelids and arthropods in the opposite direction. Evolution displays enormous ingenuity and versatility in iterating a set of common genes and developmental pathways along so many various routes of ecology and modes of life. But our brotherhood and sharing, like the still waters of legend, run far deeper than we had dared to imagine. A substantial blast from the past underlies the signs of new designs.

To end on an admittedly fatuous note, devotees of B movies will remember one of the all-time classics—the original version of *The Fly,* with

Vincent Price (not that dreadful remake with Jeff Goldblum as our hybrid hero). Focus on the unforgettable last scene: the fly with the man's head lies ensnared in a spider's web, as ugly Ms. Eight-Legs moves in for the gruesome kill. In a shrill voice of fear, the fly keeps shouting, "Please help me." Finally, and mercifully (for the fly's head on the man's body has died, so the two creatures cannot be unmixed and properly reconstituted), another character throws a rock at the web, putting fly-man out of his misery. ("They shoot horses, don't they?") Perhaps at this crucial moment in the next remake, the rock-wielding mercy killer can offer some zoological advice instead: "Turn over and be a man."

VI

DIFFERENT PERCEPTIONS OF COMMON TRUTHS

1 8

WAR OF THE WORLDVIEWS

A YEARNING FOR THE "GOOD OLD DAYS" INFECTS US ALL, EVEN THOUGH such times never existed outside our reveries. The nostalgic longing may be universal, but modes of expression vary by culture and social class. We all know the stereotypes. Plebeian Pete wishes that he could still smoke, drink, and eat red meat without raising eyebrows; while Patrician Percival laments that he just can't find dependable servants these days.

Stereotypes work by unfair exaggeration, to be sure, but they often build upon a kernel of reality. So consider this statement written in 1906 by a true Patrician Percival: "The latter minister to the former with unconscious service all the time, and with no more arrogant independence than do our domestics generally nowadays." They don't make 'em more patrician than Percival Lowell, brother of Harvard President A. Lawrence Lowell, and of poet Amy Lowell—and scion of Boston's great family in the celebrated ditty:

And this is good old Boston
The home of the bean and the cod,
Where the Lowells talk to the Cabots
And the Cabots talk only to God.

In science, as in so many other human endeavors, you don't have to be rich to succeed; but, oh my, it sure doesn't hurt either. Charles Darwin inherited a considerable fortune and then increased his stake by shrewd investments. He well understood the intellectual benefits thus acquired, primarily in freedom and time. Darwin wrote in his autobiography: "I have had ample leisure from not having to earn my own bread." But Alfred Russel Wallace, the codiscoverer of natural selection, grew up penniless, began his professional life as a schoolteacher, and always lived frugally by his wits as a writer and collector. He probably matched Darwin in intelligence, but never had the time for sustained theorizing and experiment.

Percival Lowell (1855–1916) spent his youthful *Wanderjahre* on several grand tours of Asia, leading to books with such representative titles as *The Soul of the Far East* and *Occult Japan*. He then decided to devote his life to astronomy, and began with an ultimate bang (for mucho bucks) by building a private observatory in Flagstaff, Arizona. There he did much useful work, including the prediction of a planet beyond Neptune, eventually found at his observatory by Clyde Tombaugh, and named Pluto in 1930.

But a lifetime of good work can be swallowed by one unforgettable error. Such a fate seems especially unfair when the understandable lapse of a moment erases the memory of a fine career (Bill Buckner's gimpy legs, or Pee Wee Herman's harmless impropriety). But when the error represents an *idée fixe,* relentlessly pursued over years of research and volumes of writing, then the promoter has built his own coffin. At least Percival Lowell fell before a grand enemy, the god of war himself—the planet Mars.

In the late 1870s, the Italian astronomer Schiaparelli had described the Martian surface as crisscrossed by long, thin, and straight features that he called *canali,* meaning "channels" in Italian (with no attribution of causality), not "canals" (with implications of construction by sentient beings). Lowell fell under the spell of these nonexistent phenomena, and spent the rest of his career in ever more elaborate attempts to map and interpret "these lines [that] run for thousands of miles in an unswerving direction, as far relatively as from London to Bombay, and as far actually as from Boston to San Francisco." (All quotations come from Lowell's major book on the subject: *Mars and Its Canals.)*

Lowell eventually decided that the lines must be true canals, and he developed an ever more elaborate and poignant interpretation. He viewed Mars as a once-verdant world now drying up, with polar ice caps as the only remaining source of substantial water. The canals, he decided, must represent a planetary system of irrigation, built by higher (or at least highly cooperative) beings in a last-ditch effort to funnel spring meltwaters of the ice caps to a parched and more equatorial civilization.

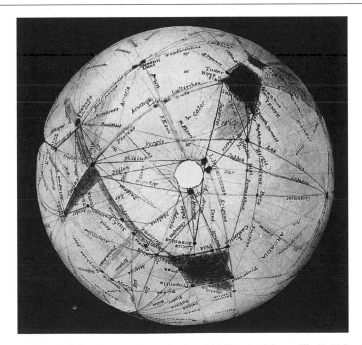

A map of the illusory Martian canals, as depicted in Percival Lowell's 1906 book, Mars and Its Canals.

As I have often emphasized in these essays, the study of error provides a particularly fruitful pathway to understanding human thought. Truth just is, but error must have reasons. If Mars had canals, then Lowell becomes an accurate observer. But a robotic photographer, insentient and without motive, might have done even better. However, since Mars does not have canals, we must ask how Lowell could have deluded himself so mightily—and the answers must embody instructive reasons and motivations. In this essay, I shall not document how Lowell decided that the canals existed, but shall concentrate instead on the logic of his argument for interpreting those supposed structures as products of a higher civilization. I choose this focus because Lowell's central error persists as a major impediment to understanding both evolution in general, and several key issues in speculations about extraterrestrial life. More immediately, the same error underlies the major public misunderstanding inspired by claims first raised in August 1996 for fossil evidence of life in a Martian meteorite.

Lowell begins his case with a false argument for extensive vegetation on Mars, an inference from supposedly seasonal changes in coloration over large portions of the Martian surface. Lowell regarded "the existence of vegetation on the planet as the only rational explanation of the dark markings there, considered not simply on the score of their appearance momentarily, but judged by the changes that appearance undergoes at successive seasons of the Martian year."

Lowell's next crucial inference inspired the complaint about servants quoted at the outset of this essay. He argues that the existence of an extensive flora implies a corresponding fauna of complex animals as well:

> Important as a conclusion this is no less pregnant as a premise. For the assurance that plant life exists on Mars leads to a further step . . . It introduces us at once to the probability of life there of a higher and more immediately appealing kind, not with the vagueness of general analogy, but with the definiteness of specific deduction. For the presence of a flora is itself ground for suspecting a fauna.

Lowell trots out all the familiar examples of interdependency between plants and animals, including pollination by insects and "preparation" of soil by earthworms—though he never seems to realize that particular cases of evolved interaction on an earth already inhabited by animals need not imply a necessary and universal linkage. After all, plants could evolve first, and animals never follow. (Lowell's forced metaphor about arrogant servants refers to animals who, in their haughty assumptions about superiority, don't even realize that they minister to plants in return for well-known service in the other direction.)

Having established (to his satisfaction) that animals must inhabit Mars, Lowell asks what level of complexity these animals must have reached, never doubting that life, once begun, must evolve to higher and higher states: "Once started," Lowell writes, "life, as paleontology shows, develops along both the floral and the faunal lines side by side, taking on complexity with time."

Lowell locates the mechanism of evolutionary advance in adaptive necessities imposed by a cooling planet. In a planet's hot youth, the simplest forms will flourish in a tropical bounty. But steady cooling requires greater organic complexity to weather increasingly harder times. When the going gets tough, to cite the current cliché, the tough get going:

> It [life] begins as soon as secular cooling has condensed water vapor to its liquid state; chromacea and confervae [unicellular plants and animals in the terminology of Lowell's time] coming into being high up toward the boiling-point. Then, with lowering temperature come the seaweeds and the rhizopods, then the land plants and the lunged vertebrates. Hand in hand the fauna and flora climb to more intricate perfecting, life rising as temperature lowers.

Mars, at a greater distance from the sun and a smaller size than Earth, must have cooled further than our planet. Martian animal life must therefore be more advanced than *Homo sapiens*. For all our vaunted wealth and technology, what have we built that a telescopic observer on Mars might recognize as a product of advanced life? The Great Wall of China? Our largest cities? Nothing, in any case, to match the scale of the Martian canals:

> Nor is this outcome [evolution to higher complexity] in any sense a circumstance accidental to the earth; it is an inevitable phase in the evolution of organisms. As the organism develops brain it is able to circumvent the adversities of condition; and by overcoming more pronounced inhospitality of environment not only to survive but spread. Evidence of this thought will be stamped more and more visibly upon the face of its habitat. On earth, for all our pride of intellect, we have not yet progressed very far from the lowly state that leaves no records of itself.

In particular, the increasing harshness of Martian environments implies a mental response from evolving life:

> In an aging world where the conditions of life have grown more difficult, mentality must characterize more and more of its beings in order for them to survive, and would in consequence tend to be evolved. To find, therefore, upon Mars highly intelligent life is what the planet's state would lead one to expect.

Put together all these conjectures and inferences—that Mars has vegetation; that plants imply animals; that both plant and animal life must progress to greater complexity; that planetary cooling and increasingly more challenging environments inspire evolutionary progress; that Mars has cooled further than Earth; that beings more advanced than *Homo sapiens* must therefore inhabit Mars—and the supposed canals cry out for interpretation as technological devices for husbanding the depleting resources of a cooling and drying planet.

Schiaparelli's "channels" must really represent what his Italian word conveys in English—true canals of planetary scale, built by higher creatures to tap the only available supply of water:

Dearth of water is the key to their character . . . So far as we can see the only available water is what comes from the semi-annual melting at one or the other [polar] cap of the snow accumulated there during the previous winter. Beyond this there is none except for what may be present in the air. Now, water is absolutely essential to all forms of life; no organism can exist without it. But as a planet ages, it loses its oceans . . . and gradually its whole water supply. Life upon its surface is confronted by a growing scarcity . . . [The canals are], then, a system whose end and aim is the tapping of the snow-cap for the water there semiannually let loose; then to distribute it over the planet's face . . . From the fact, therefore, that the reticulated canal system is an elaborate entity embracing the whole planet from one pole to the other, we have not only proof of the world-wide sagacity of its builders, but a very suggestive side-light, to the fact that only a universal necessity such as water could well be its underlying cause.

We should admire, Lowell suggests, not only the high intelligence that could build such a system, but also the superior moral qualities of beings that can cooperate (as we seem so singularly unable to do) at a planetary scale! "The first thing that is forced on us in conclusion is the necessarily intelligent and non-bellicose character of the community which could thus act as a unit throughout its globe."

Martian civilization may be doomed, despite this grand and noble planetary attempt to ward off disaster. But at least we may take courage and comfort in higher evolutionary stages thus implied for future earthly life (while we may hope for a different outcome):

One of the things that makes Mars of such transcendent interest to man is the foresight it affords of the course earthly evolution is to pursue. On our own world we are able only to study our present and our past; in Mars we are able to glimpse, in some sort, our future. Different as the course of life on the two planets undoubtedly has

345

been, the one helps, however imperfectly, to better understanding of the other.

Lowell's fanciful theory unleashed a worldwide flood of excitement and commentary, most negative (although the nonexistence of Martian canals was not conclusively established until the Mariner satellites photographed the Martian surface at close range in the 1960s. I well remember the journalistic "hook" that Lowell's old theory still provided; the popular press treated the Mariner expedition largely as a test for the existence of canals. As a young space enthusiast, I was disappointed, though not at all surprised, by the negative result!)

Alfred Russel Wallace was still alive and active when Lowell published his book on Martian canals—and still wielding his pen for a living. (Darwin and his friends, partially from guilt at their good fortune of inherited wealth, had secured an annual governmental pension for Wallace, but not nearly enough for a scholarly life free from financial worry.) Wallace, who had long been interested in the possibility of extraterrestrial life, and who had developed his own distinctive and idiosyncratic argument for earthly humanity as the universe's only example of higher intelligence in bodily form, wrote an entire book to refute Lowell's theory of canals (*Is Mars Habitable?* [London: Macmillan, 1907]).

Wallace mistakenly accepted the existence of canals and attempted to supply a purely physical explanation—as cracks "produced by the contraction of heated outward crust upon a cold, and therefore non-contracting interior." Nonetheless, he offered a devastating critique of Lowell's biological interpretation:

> The one great feature of Mars which led Mr. Lowell to adopt the view of its being inhabited by a race of highly intelligent beings . . . is that of the so-called "canals"—their straightness, their enormous length, their great abundance, and their extension over the planet's whole surface from one polar snow-cap to the other. The very immensity of this system, and its constant growth and extension

during fifteen years of persistent observations, have so completely taken possession of his mind, that, after a very hasty glance at analogous facts and possibilities, he has declared them to be "nonnatural,"—therefore to be works of art—therefore to necessitate the presence of highly intelligent beings who have designed and constructed them. This idea has colored or governed all his writings on the subject. The innumerable difficulties which it raises have been either ignored, or brushed aside on the flimsiest evidence. As examples, he never even discussed the totally inadequate water-supply for such world-wide irrigation, or the extreme irrationality of constructing so vast a canal-system the waste from which, by evaporation, when exposed to such desert conditions as he himself describes, would use up ten times the probable supply . . . The mere attempt to use open canals for such a purpose shows complete ignorance and stupidity in these alleged very superior beings; while it is certain that, long before half of them [the canals] were completed their failure to be of any use would have led any rational beings to cease constructing them.

Recent reports of fossil evidence for life in a Martian meteorite inspired me to retrieve the volumes of Lowell and Wallace from my bookshelf (where I had shelved them side by side—for history, among its many ironies, often places enemies in life into invariable positions of posthumous conjunction). These putative organisms of bacterial grade could not be more different from Lowell's wise canal builders, but I was struck by a common error that both invalidates Lowell's argument as presented above, and also underlies the fallacious main reason for public fascination with the current claim.

On August 7, 1996, NASA (the National Aeronautics and Space Administration) held a press conference to announce the publication, in the forthcoming August 16 issue of *Science* magazine, of a paper by David S. McKay and eight additional colleagues titled "Search for Past Life on Mars: Possible Relic Biogenic Activity in Martian Meteorite ALH84001." In short, these scientists argued that one of the dozen known meteorites from Mars (as reliably inferred from chemical "signatures" matching Martian atmospheric and surface conditions) contained signs of life preserved in carbonate materials deposited within cracks in the rock. These cracks presumably formed and filled on Mars some 3.6 billion years ago. (The rock was dislodged by an asteroidal impact on Mars some 15 million years ago and eventually fell on an Antarctic ice field about thirteen thousand years ago.)

The controversial evidence for life does not feature such "hard" data as a shell or a bone, but consists of chemical signals in the form of isotopic ratios and mineral precipitates often formed by biological activity (but explainable in other ways), and also of minute rod- and hairlike objects looking vaguely like the smallest of earthly bacteria, but also easily interpreted as inorganic in origin. As a betting man, I would not risk any money on the case, but neither, by any means, do I dismiss the idea. The article by McKay and colleagues is a model of caution and good sense, and their case is certainly plausible for two reasons not widely enough appreciated: first, that Mars featured appropriate conditions of running water and denser atmosphere during the first billion years of its history (when the cracks of the meteorite filled with carbonate material); second, that the Earth, at the same time and under similar conditions, *did* evolve life of bacterial grade.

The news created the greatest flourish of public interest in a scientific topic since the eruption of Mount St. Helens. Headlines captured the front page of nearly all major newspapers. *Time* magazine publicly wondered whether Mars or the nomination of Dole and Kemp at the Republican convention should command the cover. They editorialized: "One of the worst things for a news magazine is a long drought of news. Almost as challenging is being hit with two stories the same week." (They opted for the living men rather than the putative fossil bacteria, and Mars only got a corner flap of the cover, right above Jack Kemp's head.) More locally, my phone message tape filled in less than two hours with twenty-five calls from journalists.

Our leaders erupted into rapture. Bill Clinton, with a wicked sense of timing, and hoping to steal some cosmic thunder as the Republican convention opened, held a quick press conference to proclaim: "Today Rock 84001 speaks to us across all those billions of years and millions of miles." My dear friend Carl Sagan, who died just four months later, enthused from a hospital bed: "If the results are verified, it is a turning point in human history, suggesting that life exists not just on two planets in one paltry solar system but throughout this magnificent universe."

A week later in our culture of sound bites and momentary celebrity, the story disappeared from public view. Invisibility quickly descended over a discovery that many commentators had anointed as the greatest scientific revolution since Copernicus and Darwin! Lord Byron spoke in *Childe Harold* about a "schoolboy's tale, the wonder of an hour!" But Andy Warhol certainly placed his arty finger on the pulse of modern life by quartering the moment of attention to fifteen minutes, while extending the opportunity to everyone in an age without standards—from Kato Kaelin to poor Mr. Bobbitt, quickly cut off in more than one way.

But then, on second thought, maybe the story deserved no "legs," and died a proper death for lack of proof and new material to sustain the newsmongers. I would like to take an intermediate position and argue that public fascination rested largely on a false premise that guaranteed an early oblivion for the story—but that a proper formulation should sustain both interest and hope for years to come.

Lowell's logic for inferring the necessary existence of "Little Green Men" from supposed evidence of "lowly" plant life contains many errors, but none so central, or so persistent today, as the assumption that life at simplest grade must, once evolved, necessarily advance toward greater complexity and eventual consciousness. For, in such a formulation, the origin of life in *any* form implies the eventual evolution of complex creatures with consciousness, so long as planetary environments remain hospitable. Thus, Lowell argued, the origin of simple vegetation unleashes a process that must lead to canal builders. Finding the "lowest" can almost be equated with guaranteeing the "highest."

Over and over again, particularly on radio call-in programs and newspaper "man [and woman] in the street" columns, fascinated members of the public made the same mistake: if any kind of life, no matter how simple, had arisen on another world, then the evolution of consciousness must be part of a predictable natural order. (Mars became dry and frozen, thus halting the process at bacterial grade on our neighboring world. But the sequence must run to completion in many other places—for once we know that life can evolve *at all* on other planets, then Little Green Men must pervade the universe.)

The leading article in *Time* magazine began with this erroneous premise, made even worse by a false dichotomy that contrasted this supposedly inevitable scientific inference with a theological alternative:

> The discovery of evidence that life may exist elsewhere in the universe raises that most profound of all human questions: why does life exist at all? Is it simply that if enough cosmic elements slop together for enough eons, eventually a molecule will form somewhere, or many somewheres, that can replicate itself over and over until it evolves into a creature that can scratch its head? Or did an all-powerful God set in motion an unfathomable process in order to give warmth and meaning to a universe that would otherwise be cold and meaningless?

May I suggest a third alternative—by far the most probable in my view (and that of most scientists), and capable of putting the recent claim for Martian life in proper perspective. If this third viewpoint were better understood and accepted, then the putative Martian fossils would enjoy far more than fifteen minutes of false and ephemeral glory, but would foster instead a sustained search for an answer to the truly vital question that Martian life at bacterial grade should inspire.

Suppose that the simplest kind of cellular life arises as a predictable result of organic chemistry and the physics of self-organizing systems wherever planets exist with the right constituents and conditions—undoubtedly a common occurrence in our vast universe. But suppose, in addition, that no predictable directions exist for life's later development from these basic beginnings.

Evolving life must experience a vast range of possibilities, based on environmental histories so unpredictable that no realized route—the pathway to consciousness in the form of *Homo sapiens* or Little Green Men, for example—can be construed as a highway to heaven, but must be viewed as a tortuous track rutted with uncountable obstacles and festooned with innumerable alternative branches. Any reasonably precise repetition of our earthly route on another planet therefore becomes wildly improbable even in a trillion cases. (Since the universe must contain millions of appropriate planets, consciousness in some form—but not with the paired eyes and limbs, and the brain built of neurons in the only example we know—may evolve frequently. But if only one origin of life in a million ever leads to consciousness, then Martian bacteria most emphatically do not imply Little Green Men.)

In other words, I think that we have traditionally made the wrong division in a sequence of three steps: an appropriate planet without life (1), the origin of simplest cellular life at bacterial grade (2), and the evolution of consciousness (3). The traditional view, based on our arrogant assumption that life reaches a necessary apotheosis in creatures like us, assumes a wondrous specialness for life of any kind, and an inevitable evolution toward consciousness thereafter. Thus, the transition from 1 to 2 must be rare and onerous, but the passage from 2 to 3 easy and predictable.

Only under this false view can I understand the thrill felt by so many Earthlings at the recent report of Martian fossils. These people jumped to the false conclusion that a reported step 2, read as a near miracle of improbable advance from step 1, required nothing more than ample time to reach a fully predictable step 3—so that finding bacteria becomes tantamount to positing Little Green Men (a consummation never reached on Mars only because conditions changed, water disappeared, and "ample time" therefore failed to accrue). The only real difference between this common view and Lowell's canals lies in our improved knowledge of Martian geological history. Lowell thought that Martian conditions, while constantly deteriorating, had remained sufficiently hospitable to permit a predictable passage to step 3—while we now know that Mars dried up far earlier, with life still caught at step 2.

But I regard this division of the three steps as deeply erroneous, and based only on our prejudice for regarding the origin of life as special, with consciousness as a guaranteed climax thereafter. Surely the easy and constantly repeated passage lies between steps 1 and 2—representing the ordinary operation of physics and chemistry under appropriate conditions—while a transition from step 2 to step 3 faces the overwhelming improbability of any particular historical path among millions of equally attainable alternatives. Life of bacterial grade may arise almost everywhere, and then usually proceed nowhere in particular, if anywhere at all—a perfectly splendid outcome since bacteria dominate nearly all environments of life on earth even today. We now live, as Earth always has (see my 1996 book *Full House*), in an Age of Bacteria. These simplest organisms will dominate our planet (if conditions remain hospitable for life at all) until the sun explodes. During our current, and undoubtedly brief, geological moment, they watch with appropriate amusement as we strut and fret our hour upon the stage. For we are, to them, only transient and delectable islands ripe for potential exploitation.

If we could make this readjustment to view *Homo sapiens* as an ultimate in oddball rarity, and life at bacterial grade as the common expression of a universal phenomenon, then we could finally ask the truly fascinating question raised by the prospect of Martian fossils. If life originates as a general property of the material universe under certain conditions (probably often realized), then how much can the basic structure and constitution of life vary from place to independent place? We simply cannot answer this question from the only "sample" we know—life on earth—and for an interesting reason arising from the core of scientific method.

All life on earth—everything from bacteria to mushrooms to hippos—shares an astonishing range of detailed biochemical similarities, including the structure of heredity in DNA and RNA, and the universal use of ATP as an energy-storing compound. Two possible scenarios, with markedly different implications for the nature of life, might explain these regularities: either all earthly life shares these features because no other chemistry can work, or these similarities only record the common descent of all organisms on earth from a single origin that happened to feature this chemistry as one possibility among many. In the first case, life on other worlds will independently evolve the same chemistry as a sole viable choice; in the second case, other living systems may feature a wide range of alternate chemistries.

We cannot ask a more important question about the nature of life. But, ironically, we also cannot begin to answer this question with the data now at our disposal. Above all, experimental science requires repetition to test the predictability of outcomes. If a phenomenon happens only once, we simply cannot know whether the properties we observe must exist as we find them, or whether other "replays" might yield markedly different results.

Unfortunately, all life on earth—the only life we know—represents, for all its current variety, the results of a *single experiment,* for every earthly species evolved from the common ancestry of a single origin. We desperately need a *repetition* of the experiment (several would be even better, but let's not be greedy!) in order to make a judgment.

Mars represents our first real hope for a *second experiment*—the sine qua non for any proper answer to the question of questions. Unless earthly and Martian life share a single origin by seeding from a common source—an obvious possibility if Martian fossils can reach Earth by meteoritic impact!—then any life on Mars fills the holy grail of our ultimately precious second experiment.

Ancient Martian fossils will not yield the required evidence, for we need living matter with intact biochemistry, ripe for reading either as DNA, or as a workable alternative as yet unimagined by students of earthly life.

The Martian surface may now be cold, dry, and dead. But, on our planet, bacteria can live in pore spaces within rocks several miles below the earth's surface, so long as water percolates through. A similar subterranean environment on Mars may still feature water in liquid form. Thus, if life at

A close-up of the Martian surface taken by NASA's Viking Orbiter I space probe in 1980.

bacterial grade ever evolved on Mars, these organisms almost surely disappeared from the Martian surface long ago, but may still live within the more hospitable environment of subsurface rocks. The putative fossils from Mars provide our greatest reason for hope that the second experiment still lies buried, but very much alive, beneath the surface of our sister planet.

So let us send forth our robots, and perhaps (eventually) even our persons, to look, find, and return—for this experiment can be done! Forget those Little Green Men, those nonexistent canal builders, those fantasies recently unleashed under the false belief that ancient bacteria imply the eventual evolution of consciousness. The simplest life may pervade the cosmos, and a second independent sample may answer the riddle of the ages. Let us use our distinct and *oddball* intelligence to track down any direct evidence for the range of life's *universal* structure. The next step from our sun—the most accessible of all other planets—may yield a ready answer. The host of the cosmic bacterial manifold, the dominant beings who mocked Lowell for thinking that his kind rather than their kind might pervade the universe, may then smile with satisfaction and say, "So you finally understand; well done, thou good and faithful servant."

19

TRIUMPH OF THE

ROOT-HEADS

I AM NOT MUCH OF A BETTING MAN. FOR ME, A MAN O' WAR IS AN OLD British fighting ship, and a Native Dancer inhabits Tahiti, wears grass skirts, and gyrates on the beach for Fletcher Christian and Captain Bligh in various Hollywood versions of *Mutiny on the Bounty*. Nonetheless, if compelled to put up or shut up, I would make an unconventional wager on the controversial subject of progress in evolution.

In our culture's focal misunderstanding of evolution, most people assume that trends to increasing complexity through time must impart a primary and predictable direction to the history of life. But Darwinian natural selection only yields adaptation to changing local environments, and better function in an immediate habitat might just as well be achieved by greater simplicity in form and behavior as by ever-increasing complexity. Thus, one might predict that cases of evolutionary simplification will be just about as common as increases in complexity.

But I would be tempted to bet on culture's underdog, and to suspect that examples of simplification might actually hold a small overall edge. I hazard this unconventional proposition because a common lifestyle assumed by tens to hundreds of thousands of animal species—namely, parasitism—usually involves evolutionary simplification of adult form in comparison with free-living ancestors. Since I know of no comparable phenomenon that could supply a countervailing bias for complexity, a compendium of all cases might produce a majority for simplification—as natural selection in free-living forms imparts no bias in either direction, while parasitism gives a clear edge to simplification.

I regard this argument as impeccable—in its own restricted way. But nature scorns such crimping limits imposed by frailties of human cognition upon her wonderful and multifarious variety. This argument about parasites only works under the aegis of another bias almost as serious as our equation of evolution with progress: our prejudice for regarding adult anatomy as *the* organism, and our failure to consider entire life cycles and complexities of physiological function.

Consider one of the standard "laments" or "stories of wonder" in conventional tales of natural history: the mayfly that lives but a single day (a sadness even recorded in the technical name for this biological group—Ephemoptera). Yes, the adult fly may enjoy only one moment in the sun, but we should honor the entire life cycle and recognize that the larvae, or juvenile stages, live and develop for months. Larvae are not mere preparations for a brief adulthood. We might better read the entire life cycle as a division of labor, with larvae as feeding and growing stages, and the adult as a short-lived reproductive machine. In this sense, we could well view the adult fly's day as the larva's clever and transient device for making a new generation of truly fundamental feeders—the insect equivalent of Butler's famous quip that a chicken is merely the egg's way of making another egg.

This essay treats the most celebrated story of extreme simplification in an adult parasite—in the interests of illuminating, reconciling, and, perhaps, even resolving two major biases that have so hindered our understanding of natural history: the misequation of evolution with progress, and the undervaluing of an organism by considering only its adult form and not the entire life cycle.

The adult of *Sacculina,* the standard representative of a larger group with some two hundred species, the Rhizocephala, could hardly be more different from its barnacle ancestors—or more simplified in anatomy and appearance. The two names accurately record this dramatic evolutionary change—for *Sacculina* is a Latin "little sac," while *Rhizocephala* is a Greek "root-head." As we shall see, the rhizocephalans are clearly barnacles by ancestry, but the adult preserves not a hint of this crustacean past. Rhizocephalans are parasites upon other crustaceans, and nearly all infest decapods (crabs and their relatives). The adult consists of two parts with names that (in a refreshing change from usual practice) almost count as vernacular expressions, rather than jargon. From the outside, a human observer sees only a formless sac (called the *externa*) attached to the underside of the crab's abdomen. The sac is little more than a reproductive device, containing the ovary and a passageway for introduction of males and their sperm. The externa contains no other differentiated parts—no appendages, no sense organs, no digestive tract, and no sign of segmentation at all. The

*E. Ray Lankester's figure of the "degenerate" adult Sac-*culina, *showing the externa (sac) and interna (roots).*

fertilized eggs develop within the externa (which then operates as a brood pouch).

But how can the externa function without any evident source of nutrition? Closer examination reveals a stalk that pierces the crab's abdomen and connects the externa to an elaborate network of roots (called the *interna*). These roots may pervade the entire body of the crab. They penetrate through the hemocoelic spaces (the analogs of blood vessels) and invest many of the crab's internal organs. They provide nutrition to the parasite by absorption from the crab's vital fluids. In some species, roots are restricted to the abdomen, but in *Sacculina* they may run through the entire body, right to the ends of the appendages. (This system is not so grisly—in inappropriate human terms—as first glance might suggest. The parasite does not devour the host, but rather maintains the crab as a "life support" system.) The name *Sacculina* (for the most common genus) honors the externa, while the designation of the entire group—Rhizocephala, or root-head—recognizes the interna.

These barnacle parasites have been known to zoologists since the 1780s (though the first recorder, correctly observing the release of crustacean larvae from the externa, misinterpreted the sac as an organ of the crab, induced by the parasite much as some insect larvae can commandeer a plant to grow a protective gall). Ever since this early discovery, rhizocephalans have played a classic role in conventional natural history as the standard example of maximal degeneration in parasites. Many of the foremost zoologists of Darwin's generation highlighted *Sacculina* as one of evolution's primary marvels.

The German biologist Fritz Müller wrote a famous book in 1863 that provided Darwin with crucial early support. Müller's book deals almost entirely with the anatomy of crustaceans, but bears the general title *Für Darwin (For Darwin)*. Müller cited *Sacculina,* and its undoubted relationship with free-living barnacles, as a primary example of "retrogressive metamorphosis" in evolution. He referred to this genus as "these *ne plus ultras* in the series of retrogressively metamorphosed Crustacea," and he wrote of their limited activity:

> The only manifestations of life which persist . . . are powerful contractions of the roots and an alternate expansion and contraction of the body, in consequence of which water flows into the brood-cavity, and is again expelled through a wide orifice.

E. Ray Lankester (1847–1929), later director of the Natural History Division of the British Museum, published a famous essay in 1880 titled *Degeneration: A Chapter in Darwinism*. He defined degeneration as "a loss of organization making the descendant far simpler or lower in structure than its ancestor," using *Sacculina* as a primary example. Lankester described the barnacle parasite as "a mere sac, absorbing nourishment and laying eggs."

Yves Delage (1854–1920), one of France's finest natural historians and a patriotic Lamarckian, published a major empirical study on *Sacculina* in 1884. He referred to the genus as "this singular parasite, reduced to a sac containing the genital organs." "*Sacculina,*" he added, "seems to be one of those beings made to chill adventurous imaginations" *(faits pour refroidir les imaginations aventureuses)*.

Thus, all major authors and experts used the Rhizocephala as primary illustrations of degeneration in the evolution of parasites (or, at least, of simplification if we wish to avoid the taint of moral opprobrium). I will not challenge this assertion for a restricted view of the adult as an external sac attached to internal roots. But I do wish to oppose the myopia of such a restriction. From a properly expanded viewpoint—and for three major reasons that I shall discuss in sequence—rhizocephalans are remarkably intricate animals, as bizarre in their elaborate uniquenesses as any creature on earth. In this expanded perspective, however, they remain as wonderfully provocative as ever—as superbly illustrative of the meaning of evolution as when Europe's greatest zoologists falsely appointed them as chief exemplars of Darwinian degeneration.

1. THE FULL LIFE CYCLE OF THE RHIZOCEPHALA. How did we ever discern the barnacle ancestry of *Sacculina*? We could now gain this information by sequencing DNA, but early-nineteenth-century zoologists correctly identified the affinity of rhizocephalans. How did they know, especially when studies of the adult externa and interna could not provide the slightest clue?

Observations of the complex life cycle in female rhizocephalans solved this zoological puzzle. (I shall discuss the growth of males later, as my third argument.) The first two phases of growth differ very little from the development of ordinary barnacles, and therefore seal the identification. The larvae exit from the externa's brood pouch as a conventional dispersal stage, common in many crustaceans, called the *nauplius*. The rhizocephalan nauplius passes through as many as four instars (molting stages) and, except for the absence of all feeding structures, looks like an ordinary crustacean nauplius, right down to the most distinctive feature of a single median eye.

I am trying to suppress my usual lateral excursions in this essay—if only because I find the main line of the story so exciting—but I cannot resist one digression for its striking illustration of science's human face. Yves Delage's 1884 monograph on *Sacculina,* undoubtedly the most important early study of rhizocephalans, runs to more than three hundred pages of dry anatomical description, devoted mainly to these early stages of the life cycle. But at several points he vents his anger at a German colleague, R. Kossmann. Delage took particular delight in exposing Kossmann's error in identifying two larval eyes. Early in his monograph, this French patriot admits the source of his venom and consequent pleasure in Kossmann's mistakes. Kossmann had previously skewered a Frenchman, a certain Monsieur Hesse, for errors in interpreting the life cycle of *Sacculina*. Delage took offense for two reasons. First of all, poor Hesse was a dedicated amateur who only took up the study of marine zoology in retirement, "at an age when so many others, in Germany as elsewhere, are only seeking to enjoy the inactivity of repose merited by their long service." Kossmann should have been more generous. But second, and impossible to forgive, Kossmann had explicitly attacked Hesse *as a Frenchman* in clear violation of the norms of science as a cooperative and international enterprise. Delage then specu-lated about Kossmann's motives and recalled his own bitter feelings at the defeat of his country in the Franco-Prussian War of 1872:

> What I cannot excuse is that this gentleman [Mr. Kossmann] expressed pleasure in seeing a scientist fall into error because that scientist is a Frenchman. This illustrates the workings of a narrow mind, and such thinking will quickly destroy the characteristic nobility of scientific discussion. But Mr. Kossmann has an excuse. Note that he wrote in 1872, at a moment when Germany was still tipsy from its recent military successes, and he just didn't have enough fortitude to resist the temptation to give the proverbial kick in the behind to the defeated.

The one-eyed nauplius only identifies rhizocephalans as crustaceans, but the next phase, the cyprid larva, occurs only in barnacles and thus spec-ifies the ancestry of the root-heads. If the nauplius acts as a waterborne dis-persal phase, the subsequent cyprid explores the substrate by crawling about on a pair of frontal appendages called antennules, securing a good spot for attachment, and then secreting cement for permanent fastening. This cement fixes most barnacles to rocks, but some species attach to whales or turtles, and one species sinks deep into whale skin to live as a near parasite. Thus, we can easily envisage the evolutionary transition from fastening to rock, to external attachment upon another animal, to internal burrowing for protection, and finally to true internal parasitism. In any case, the rhizo-cephalan cyprid functions like its barnacle counterpart and searches for an appropriate site of attachment upon a crustacean host. (Favored sites vary from species to species; some settle on the gills, others on the limbs.)

We now reach the crux of the argument in considering the curious uniqueness of rhizocephalans as defined by newly evolved stages in an intri-cate life cycle. How does the cyprid, now attached to an external part of the host, manage to get inside the host's body to become an adult root-head?

The rhizocephalan life cycle proceeds from taxonomic generality to uniqueness. The initial nauplius identified the creature as a crustacean; the subsequent cyprid proves barnacle affinities within the Crustacea. But the next phase belongs to root-heads alone.

The female cyprid, now attached to the host by its antennules, metamorphoses to a phase unique to the rhizocephalan life cycle, as discovered by Delage in 1884 and named the *kentrogon* (meaning "dart larva"). The kentrogon, smaller and simpler than the cyprid, develops a crucial and special organ—Delage's "dart" (now generally called an "injection stylet"). The kentrogon's dart functions as a hypodermic needle to inject the precursors of the adult stage into the body of the host!

This delivery system for the adult's primordium shows great diversity across the two hundred or so species of rhizocephalans. In one group, the

Yves Delage's original figure of a root-head kentrogon, or "dart larva," injecting precursors of its adult stage into the body of a crab. Note the dart, at top, piercing a piece of the crab's body.

kentrogon cements its entire ventral surface to the host. The dart then pierces the host through this ventral surface, requiring a passage through three layers—the kentrogon's cuticle, the attaching cement, and the host's cuticle. In another group, the kentrogon's ventral surface does not cement, and antennules continue to function as attachments to the host. In these forms, including the genus *Sacculina* itself, the injecting dart goes right through one of the antennules, and thence into the body of the host! A third group skips the kentrogon stage entirely; the cyprid's antennule penetrates the host and transfers the primordial cells of the adult parasite.

Yves Delage, who discovered the kentrogon and its injecting device in 1884, could not hide his amazement. He wrote:

> All these facts are so remarkable, so unexpected, so strange compared with anything known either in barnacles or anywhere in the entire animal kingdom, that readers will excuse me for providing such a thorough factual documentation.

But the next observation strikes me as even more amazing—the high-point of rhizocephalan oddity, and a near invitation to disbelief (if the data were not so firm). What constitutes the primordium of the adult parasite? What can be injected through the narrow opening of the dart's hypodermic device?

Delage, who discovered the mechanism, concluded that several cells, maintaining some organization as precursors to different tissues of the adult, entered the host. He could hardly come to grips with the concept of this much reduction separating larval and adult life. Imagine going through such complexity as nauplius, cyprid, and kentrogon—and then paring yourself down to just a few cells for a quick and hazardous transition to the adult stage. What a minimal bridge at such a crucial transition! "The *Sacculina,*" Delage wrote, "has been led to make something of a tabula rasa [blank slate] of its immediate past." Delage then groped for analogies, and could only come up with a balloonist jettisoning all conceivable excess weight upon springing a leak.

All can be explained by the necessity for the parasite to make itself very small in order to pass more easily through the narrow canal, whose dimensions are set by the orifice of the dart. [The transferred cells] are in the same condition as an aeronaut whose balloon has lost part of its gas, and who, needing to rise again at all cost, lightens his load by throwing out everything not absolutely indispensable to the integrity of his machine.

Well, Monsieur Delage, the actual situation far exceeds your own source of amazement. You were quite right; many species do transfer several cells through the dart. But other species have achieved the ultimate reduction to a single cell! The dart injects *just one cell* into the host's interior, and the two parts of the life cycle maintain their indispensable continuity by an absolutely minimal connection—as though, *within* the rhizocephalan life cycle, nature has inserted a stage analogous to the fertilized egg that establishes minimal connection *between* generations in ordinary sexual organisms.

The evidence for transfer of a single cell has been provided in recent articles by our leading contemporary student of rhizocephalans, Jens T. Høeg of the zoological institute of the University of Copenhagen. (I read about a dozen of Høeg's fascinating papers in preparing this essay, and thank him for so much information and stimulation.) In a 1985 article on the species *Lernaeodiscus porcellanae,* published in *Acta Zoologica,* Høeg documented the settlement of cyprids, formation of kentrogons, and injection of only a single cell, recognizable within the kentrogon, into the host. Høeg writes of the tenuous bridge within the kentrogon: "Because of its size and apparent lack of specialization the invasion cell stands out conspicuously against the surrounding epithelial, nerve and gland cells."

The September 14, 1995, issue of *Nature,* my original inspiration for this essay, reports an even more remarkable discovery: "A new motile, multicellular stage involved in host invasion by parasitic barnacles (Rhizocephala)," by Henrik Glenner and Jens T. Høeg. The authors found that the kentrogon of *Loxothylacus panopaei* injects a previously unknown structure into the host:

a wormlike body containing several cells enclosed in an acellular sheath. This "worm" breaks up within the host's body, and the individual cells, about twenty-five in number, then disperse separately "by alternating flexure and rotating movements." Apparently, each cell maintains the potential to develop into an entire adult parasite, though only one usually succeeds (a few crabs develop multiple externa with independent root systems inside the body).

This minimal transition helps to explain why the adult root-head shows no sign of barnacle affinities. If the adult parasite develops anew from a single transferred cell, then all architectural constraints of building an adult

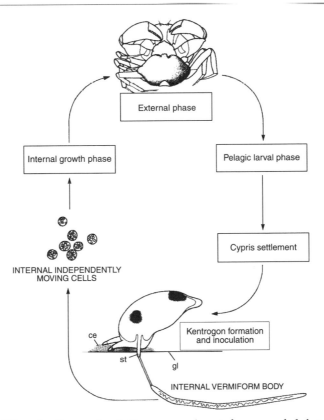

The complex life cycle of a root-head: Recent experiments have revealed that, in some species, the kentrogon injects the host crab with a motile, worm-shaped body that later splits into several independently moving cells.

from parts of a taxonomically recognizable larva have been shed. In any case, the primordial cell or cells then migrate from the site of injection, through the circulatory spaces of the host, find a site for settlement, build an internal root system, and finally emerge through the host's abdomen as a new structure bearing the charming name of "virgin externa."

Take all this—nauplius; cyprid; kentrogon; injected passage into the host's body, sometimes by a single cell; migration to a permanent site; reproduction of the rooted interna; and emergence of the externa. Stack up these stages against our own lives, even through all the Sturm und Drang of our teenage years, and which life cycle would you label as more "complex"?

2. MANIPULATING AND COMMANDEERING THE HOST. The adult parasite may look like a rooted blob, but just as the most unprepossessing humans often hide immense power beneath their ordinary appearance (as many a Hollywood "toughie" has discovered to his great hurt and sorrow), beware of equating ugly wartiness with benign simplicity. The adult rhizocephalan parasite has more tricks up its nonexistent sleeve than externa appearance would suggest.

Consider the following problem in logic as an indication of the physiological and behavioral sophistication that adult parasites must possess. We know that crabs fight back when the cyprid larvae try to settle, for potential hosts use their cleaning and grooming behaviors to remove the settling cyprids—and the great majority of potential parasites are thereby destroyed. In fact, the rapid transformation of exploring cyprid to cementing kentrogon (accomplished within ten minutes in some species), the low and hunkering shape of the kentrogon, and its firm cementation to the host in many species have all been interpreted—quite correctly in my view—as active adaptations by the parasite to vigorous counterattacks by potential hosts.

But when the virgin externa pokes through the abdomen and lies flush against the crab's underside, all "fight" has evaporated from the host. The crab still possesses an active cleaning response, but makes no attempt to remove the externa. Why not? What has happened to the crab? In a remarkable paper published in 1981 in the *Journal of Crustacean Biology,* authors Larry E. Ritchie and Jens T. Høeg answer these questions in studying the root-head species *Lernaeodiscus porcellanae:*

> The parasite returns to the surface as the externa. What keeps the host from recognizing it as foreign or "parasite" and destroying it, since the cleaning behavior is still available? When an externa appears on the surface of the host, it must either be in a position or of a form that cannot be removed by the host, or it must be perceived as "self" and not harmed in any way.

Since the externa could presumably be reached and removed, the second and more interesting alternative probably applies. In other words, the parasite has somehow evolved to turn off the host's defenses, presumably by disarming the crab's immune response with some chemical trickery that fools the host into accepting the parasite as part of itself. The authors continue:

> The evolution of host control, probably through some form of hormonal action, represents the ultimate counterdefensive adaptation of the Rhizocephala, for it nullifies the host's defense system . . . Once host control is achieved, the host is in the absolute service of the parasite.

The phrase "absolute service" may sound extreme, but a compendium of parasitic devices for usurpation, takeover, and domination of the host can only elicit an eerie feeling of almost macabre respect for the unparalleled thoroughness (and cleverness) of parasitic management!

First of all, the adult parasite castrates the host, not by directly eating the gonadal tissue (as in most cases of "parasitic castration," a common phenomenon in this grisly world), but by some unknown mechanism probably involving penetration of the interna's roots around and into the crab's nervous system. In *Sacculina* (but not in most other rhizocephalans), the parasite also cuts off the host's molting cycle, and the crab never again sheds its

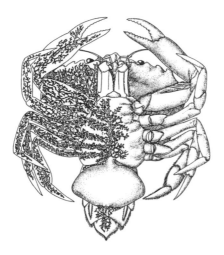

The "roots" of the adult Sacculina *pervade the body of a crab, nourishing the parasite and castrating—but not killing—the host.*

outer shell (an obvious benefit to the externa, which can easily be dislodged by molting).

Lernaeodiscus porcellanae turns control of the host into a fine art. After castration by the parasite, male crabs develop female characteristics in both anatomy and behavior, while females become even more feminized. The emerging externa then takes the same form and position as the crab's own egg mass (in normally developing uncastrated females)—attached to the underside of the abdomen. The crabs—both male and female (for both sexes are feminized by the parasite)—then treat the externa as their own brood. In other words, the parasite usurps all the complex care normally invested in the crab's own progeny. Crabs ventilate the externa by waving their abdomens; they actively (and carefully) groom the externa with their cleaning limbs. Moreover, Ritchie and Høeg proved that this behavior may be indispensable for the externa's survival—for when they removed the cleaning limbs from a parasitized crab, "the externa soon became fouled and necrotic." Finally, the "simple" root-head even fools the crab into treating the release of parasite larvae from the externa as the discharge of her own fertilized eggs! Ritchie and Høeg write:

When it is time for the parasite to release its larvae, the host assists by performing customary spawning behavior. Normally cryptic, [the crab] climbs out from under the rock, elevates the body on tip-toes, and then lowers and raises the abdomen in a waving action. Simultaneously, the parasite expels its nauplii into the current generated by the host.

In short, rhizocephalans are the cuckoos of the marine invertebrate world—laying their eggs in another species's "nest," mimicking the host's own eggs (similarity of the externa to the crab's egg mass), and then eliciting parental care from the host. But rhizocephalans are even more thorough, for they always castrate their host, while only some cuckoos kill the legitimate nestlings of their foster parents.

In short, the root-head turns the crab into a Darwinian cipher, a feeding machine working entirely in the parasite's service. The castrated crab can make no contribution to its own evolutionary history; its "Darwinian fitness" has become flat zero. All feeding and growth now work in the evolutionary interest of the root-head, which continues to reproduce at a prodigious rate, entirely at the crab's expense—as the interna's roots drain the crab's nutrition. But ever so carefully, for the parasite must maintain the crab in constant and perfect servitude—not draining the host enough to kill this golden goose, but not letting the crab do anything for its own Darwinian benefit, either.

Root-heads can maintain this delicate balance for a long time. Ritchie and Høeg kept infested crabs for two years in the laboratory—with the parasites showing no ill effects and reproducing all the time. Moreover, the root-head can produce prodigious numbers of larvae—all supported by the crab's feeding. In a 1984 article on *Sacculina carcini*, Jørgen Lützen found that a single externa, during a breeding season lasting from mid-July to October, can produce up to six batches of eggs, with an average of 200,000 per clutch—for a total of more than a million eggs per season. Complex indeed—and devilishly effective. If I were a conscious rhizocephalan, I would adopt this motto: Don't call me a simple sac with roots.

3. WHAT ABOUT MALES? (OR, FURTHER COMPLEXITY IN THE ROOT-HEAD LIFE CYCLE). These first two categories of complexity only consider the female root-head. Delage and all early students of rhizocephalans regarded the externa as a hermaphrodite, with both male and female organs. But the externa forms part of the adult female only. Male rhizocephalans were not well documented until the 1960s, when a group of Japanese zoologists finally worked out the full sexual system of rhizocephalans, and recognized the true nature of males.

The male life cycle differs in a striking way from the development of females—another testimony to rhizocephalan complexity when we consider full biology rather than adult anatomy alone. The beginning stages of nauplius and cyprid differ little between the sexes. But whereas the female cyprid settles on a crab to begin the stage of internal penetration, the male cyprid alights instead on the female externa. In *Sacculina* and close relatives, the virgin externa contains no opening. But this initial externa soon molts to a second stage containing an orifice known as the "mantle aperture." This opening leads into two passageways known as "cell receptacles."

Successful male cyprids settle on the externa's aperture. A unique male stage, called the *trichogon,* then forms within the cyprid. The trichogon, clearly the homolog of the female kentrogon, but much simpler in form, has no muscles, appendages, nervous tissue, or sense organs. The trichogon looks like a small mass of undifferentiated cells surrounded by a cuticle covered with small spines (the name means "hairy larva"). The trichogon passes through the antennule of the cyprid, into the aperture of the externa, and down the passageway of the cell receptacle. (Two trichogons may successfully enter the externa, one in each receptacle.) The trichogon then sloughs the spiny cuticle and lodges as a small group of cells at the end of the passageway. (Other rhizocephalans form no trichogon; the male cell mass must then be injected through the cyprid antennule right into the body of the externa.) These tiny male cell masses become sources for the production of sperm.

These facts may not warm the hearts of superannuated macho blusterers among humans, but male root-heads end up as tiny dwarfs, injected into the body of a vastly larger female, and finally coming to rest as a small mass of loosely connected cells deep inside the externa. Biologists refer to such males as hyperparasites—for they are parasitic upon a parasite. The female parasitizes a crab, but the tiny male depends entirely upon the female for nourishment. The male cells, permanently enclosed deep within the body of a relatively enormous and protective female (an odd kind of Freudian fantasy even for human males, I suppose), then spend their days producing sperm in synchrony with egg production by the externa.

In summary, the intricate and different life cycles of both male and female root-heads, and the great behavioral sophistication shown by the female in reconfiguring a host crab as a support system, all underscore the myopia of our conventional wisdom in regarding rhizocephalans as degenerate parasites because the adult anatomy of internal roots and external sac seems so simple.

This reassessment of root-heads forces me to revise my initial take on the lessons of parasites for correcting the bias of equating evolution with progress—for I can no longer hold that parasitic degeneration argues for a slight preponderance of simplification over complexification among evolutionary trends. But this correction leads to an even better argument against predictable progress—one that also takes us back to the roots of our intellectual heritage in Darwin's ideas.

In his famous 1880 essay *Degeneration: A Chapter in Darwinism*, E. Ray Lankester correctly identified belief in progress as the principal inference falsely drawn from Darwin's theory of natural selection. Lankester wrote:

> Naturalists have hitherto assumed that the process of natural selection and survival of the fittest has invariably acted so as either to improve and elaborate the structure of all the organisms subject to

371

it, or else has left them unchanged, exactly fitted to their conditions, maintained as it were in a state of balance. It has been held that there have been some six or seven great lines of descent . . . and that along each of these lines there has been always and continuously a progress—a change in the direction of greater elaboration.

Lankester then cited supposed cases of degeneration, including root-heads as a primary example, to prove that natural selection does not guarantee such progress. "Degeneration," he wrote, "may be defined as a gradual change of structure in which the organism becomes adapted to less varied and less complex conditions of life." Lankester, in other words, remains true to Darwin's deeper principle that natural selection leads only to local adaptation, not to global progress. He correctly states that simplified conditions of life might lead, by natural selection, to less complex anatomies—and that these simplified descendants would be just as well adapted to their habitats as more complex ancestors to previously more elaborate modes of life. But Lankester erred in regarding root-heads as degenerate forms properly responding to simplified conditions. An enlarged view of the entire root-head life cycle reveals great complexity and corresponding adaptation through several intricate phases of growth. The simple adult sac of the female externa tells only a tiny part of a fully elaborate tale.

Rhizocephalans, instead, provide a superb example of Darwin's genuine principle—the production of appropriate local adaptation by natural selection. Rhizocephalans are phenomenally well suited to their complex conditions of life. But in evolving their unique specializations, rhizocephalans did not become better (or worse) than any close relative. Are root-heads better than barnacles because they live in crabs rather than on rocks? Are they worse than barnacles because the adult female looks like a bag, rather than a set of gills enclosed in a complex shell? Is a crab better than a barnacle? Do we prefer seahorses over marlins, bats over aardvarks? Such questions are foolish and diversionary.

Natural selection can only adapt each creature to its own local conditions—and such a mechanism therefore cannot serve as a rationale for our oldest and most pernicious prejudice of progress. Rhizocephalans derail the bias of progress not because they are degenerate, but because they are so well adapted and uniquely specialized to their own intricate series of lifetime environments—and how can we possibly rank all the disparate uniquenesses of the animal kingdom as cosmically better or worse? May this aid provided to *our* poor benighted intellects—not only *their* undoubted success in commandeering crabs for Darwinian advantage—represent the triumph of the root-heads.

2 0

CAN WE TRULY KNOW

SLOTH AND RAPACITY?

THE CLASSIC GENERALIZED STATEMENT OF A COMMON PROBLEM IN INTELlectual and practical life may be found in Tennyson's lament that his dearest (and deceased) friend Arthur Hallam seemed so close in loving memory, yet so unreachable in actuality:

> *He seems so near, and yet so far.*

The classic particularized statement of the same problem describes Coleridge's Ancient Mariner, adrift in an utterly unusable but completely enveloping bounty:

> *Water, water, every where,*
> *Nor any drop to drink.*

The common experience of being so close that you can almost touch, yet so utterly distant by any available way of knowing, provides a decidedly mixed blessing—as both a primary frustration of daily life and a major prod to scientific advance. My favorite example has a largely happy ending still vigorously in progress; consider how much of medicine's sorry history of so little advance over so many centuries (until very recently) arose primarily from a diagnostic problem involving only an inch or two. The trouble requiring visualization lies just below the opaque covering of our skin. Untold millions (probably billions) of premature, and often painful, deaths have occurred because no one could see a developing tumor or an internal source of infection. The old surgeons could do little more than cut it off or (on occasion) take it out—where "it" refers to something quite large (a limb, for example), removed *in toto* because a small and local lesion could not be pinpointed. Imagine, then, the triumphant benevolence of a host of inventions from X rays to CT scans to MRIs. The ability to see an inch or two inside has revolutionized our lives and greatly improved our prospects.

Natural historians have dedicated themselves to the noble and fascinating task of trying to understand, in the deepest way accessible to us, the amazing variety of life on our planet. The best possible procedure immediately runs into Tennyson's limit of proximity with impossibility. I go eyeball to eyeball with some other creature—and I yearn to know the essential quality of its markedly different vitality. I cry to God the Gatekeeper of scientific knowability: Give me one minute—just one minute—inside the skin of this creature. Hook me for just sixty seconds to the perceptual and conceptual apparatus of this other being—and then I will know what natural historians have sought through the ages.

But this god stays as silent as Baal, who would not answer the loud and fervent pleas of his 450 prophets, even when Elijah mocked them for the impotence of their deity. I can only look from the outside (or cut into the inside, but flesh and genes do not reveal organic totality). I am stuck with a panoply of ineluctably indirect methods—some very sophisticated to be sure. I can anatomize, experiment, and infer. I can record reams of data about behaviors and responses. But if I could *be* a beetle or a bacillus for that one precious minute—and live to tell the tale in perfect memory—then I might truly fulfill Darwin's dictum penned into an early notebook containing the first flowering of his evolutionary ideas during the late 1830s: "He who understands baboon would do more towards metaphysics than Locke."

Instead, we can only peer in from the outside, look our subject straight in the face, and wonder, ever wonder. Still, considering how far our methods must lie from the unreachable optimality of dwelling within, we have managed pretty well in a world without metempsychosis. Our indirect methods have taught us a mountain of things about horses, but if you wished to learn even more, wouldn't you rather be Whirlaway in the stretch, than interview Eddie Arcaro afterwards?

I came face to face (many times) with this old paradox during a recent trip to Costa Rica, a nation justly celebrated for its maximal attention among poor and tropical lands to the health and preservation of remaining natural environments—a position not only ethically correct, but also potentially profitable both to a nation itself, and to all of us. Two Costa Rican animals cast a particularly enigmatic stare at me, and elicited the old frustrating thought that if I could only get inside their different world for a minute, I might understand. Small mammals and insects strike us as frenetic; some reptiles and amphibians seem overcome with torpor. But we are not overwhelmed with the difference, if only because all these creatures vary their routines and paces: a squirrel can sit rigidly still, while an "immobile" frog catches insects on a lightning tongue.

But sloths move with such pervasive slowness that their entire world seems intrinsically and permanently different from ours. I would almost conjecture that a fixed slow-motion camera occupies their cranial space, and that they gauge all their movements by this markedly different clock upon the world. Do we, and most other creatures, appear to them like the Keystone Kops in movement, or the Munchkins in raised pitch? Or do our frenetic paces (compared with their stately step) constitute the only external world they know, recorded in their brains as the slothful equivalent of "objective reality"? If one of El Greco's uniformly tall and thin people stepped out of a canvas featuring no other person but himself, and then entered our world, would we all appear ridiculously squat and fat, or would he know nothing else (by virtue of a few centuries' experience with human gawkers in art galleries, but never even a glance at *confrères* on other canvases), and therefore view us as ordinary and archetypal? But sloths do know other sloths and must also perceive a differently paced external world as well. Perhaps they don't notice the difference; perhaps they are merely amused; perhaps they don't care. I would love to know.

In any case, philosophical speculation aside, I have never been so powerfully moved by a sense of pervasive difference for something so basic as a pace of life. Hanging upside down, and grasping a tree branch by all fours, sloths move along hand over hand, and so very slowly—not (apparently) for reasons of immediate caution, but in accord with their own concept of normality. They stretch out an arm to reach their leafy food with the same utter languor. The algae that grow on their hanging hairs, imparting a green tinge to the entire body, almost seem to take hold because the animal can't move away fast enough. (Yes, of course, I intend the last sentence only as a metaphor—but then I once heard that a rolling stone gathers no moss!)

At least my impressions are not idiosyncratic. Sloths seem to impact all Western observers in the same basic way. Englishmen named them with a word meaning "slow" by etymology, thus identifying sloths with one of the Seven Deadly Sins as well. Every other language that I know uses the same designation. They are *paresseux* in French, *perezoso* in Spanish, and *pigrizia* in Italian—all meaning "lazy" or "indolent." They are *ignavus* in Linnaeus's Latin, meaning the same thing—and Linnaeus formally named a sloth genus *Bradypus,* meaning "flow-foot" in Greek. As I stood watching a sloth high in a tree at Manuel Antonio National Park, I heard a group of German tourists speaking about a "foul" animal. I thought that they just didn't care for the poor creature, but then I remembered that *faul* is German for "lazy," and that a sloth *auf Deutsch* is *Faultier.*

But how slow is slow? (As I wrote in introducing the topic of this essay, we can at least experiment and accumulate outward data in lieu of our real desire to get inside another animal's head.) Early sources heaped the calumny of exaggeration upon a reality already genuine enough. Nehemiah

Grew, the first scientist (according to the *Oxford English Dictionary*) to call them by their common English name, wrote in 1681, in his catalog of specimens owned by the Royal Society of London: "The sloath . . . An animal of so slow a motion, that he will be three or four days, at least, in climbing up and coming down a tree." Linnaeus, when formally naming this creature in his mid-eighteenth-century *Systema naturae,* wrote: *"tardissime et aegre incedit, vix uno die 50 passus"* ("he moves most slowly and reluctantly, scarcely managing 50 paces in a day").

Their step is, in fact, a bit brisker, though nothing to challenge Aesop's tortoise. In the standard book on the subject, *Function and Form in the Sloth,* M. Goffert begins his chapter on "motor activity" by writing: "Sloths sleep or rest about twenty hours a day, performing perhaps no more than 10 percent of the work of a higher mammal of the same size." Goffert then summarizes a number of careful studies devoted to measuring the speed of sloths. Their movement along a horizontal pole (a good experimental surrogate for their favored tree branches in nature) averages a stately 0.1 to 0.3 miles per hour, with maximal acceleration to a sprightly 1.0 miles per hour.

Since sloths are so evidently well adapted to motion upside down along tree branches, we should not be surprised that their infrequent right-side-up progression on the ground should be so painfully inefficient. With front legs longer than hind limbs, and with permanently curved digits that hook well to branches but permit only hobbling motion on the ground, sloths cannot manage more than 0.1 to 0.2 miles per hour on terra firma—scarcely enough to outrun a pursuing jaguar.

Several aspects of sloth anatomy and physiology correlate with their extreme slowness. Studies of contraction time show that, in Goffart's words, "the muscles of the fastest genus of sloths were thus four to six times slower than their homologues in the cat." Sloths also maintain a lower and more variable body temperature than almost any other mammal—a fact of undoubted relevance to their slow pace of life. Most mammals hold their steady body temperature just a bit below 100°F (as in our "standard" of 98.6°). Monotremes and marsupials, the egg-laying and pouched mammals of Australia and a few other places, operate at a considerably lower level; the duck-billed platypus, for example, maintains its minimally warm-blooded body at about 85°F.

Sloths belong to the exclusively New World mammalian order Edentata, including armadillos and three genera of South and Central American anteaters. Edentates maintain the lowest body temperatures among placental mammals. For example, two species of the sloth genus *Bradypus* varied between 82° and 90°F throughout the day, depending upon the outside temperature.

Yet, for all these attempts to approach the sloth's inner reality with our best inferences from outward data, we have failed badly (and for the usual reason of inability to overcome our self-centered view), at least in popular presentations. From the name that serves as their definition and incubus, to our constant emphasis on their slowness, stupidity, and dull daily routines, we have conveyed an image of sloths as very low mammals doing very little of interest very high in the trees. This tradition began with a remarkable characterization by the great French naturalist Georges Buffon in his classic eighteenth-century compendium, the many-volumed *Histoire naturelle*. Buffon held sloths in maximal contempt among mammals, and expressed his derision (in his usual elegant prose) by explicit comparison with human abilities, rather than by any attempt to grasp the sloth's own world of opportunities and dangers. Buffon wrote (my translation):

> Whereas nature appears to us live, vibrant, and enthusiastic in producing monkeys; so is she slow, constrained, and restricted in sloths. And we must speak more of wretchedness than laziness—more of default, deprivation, and defect in their constitution: no incisor or canine teeth, small and covered eyes, a thick and heavy jaw, flattened hair that looks like dried grass . . . legs too short, badly turned, and badly terminated . . . no separately movable digits, but two or three excessively long nails . . . Slowness, stupidity, neglect of its own body, and even habitual sadness, result from this bizarre

and neglected conformation. No weapons for attack or defense; no means of security; no resource of safety in escape; confined, not to a country, but to a tiny mote of earth—the tree under which it was born; a prisoner in the middle of great space . . . everything about them announces their misery; they are imperfect productions made by nature, which, scarcely having the ability to exist at all, can only persist for a while, and shall then be effaced from the list of beings . . . These sloths are the lowest term of existence in the order of animals with flesh and blood; one more defect would have made their existence impossible.

As if Buffon had not already heaped enough disdain upon sloths, he then argues that human misery arises from moral failures of conscious decisions, and not from inborn propensity. But only among sloths has nature decreed inherent degradation:

> The disgraced sloths are perhaps the only creatures that nature has maltreated, the only creatures that offer us an image of innate misery.

Only at the very end does Buffon pull back a bit, wonder about the sloth's own internal state (as this essay advises), and conjecture that things may not be so bad after all—for such an insensible creature might not grasp its own plight:

> If the misery resulting from lack of feeling is not the greatest of all ills, then that of these animals, although very apparent, may not be real, because they appear to feel so little: their mournful appearance, their heavy look, their indolent insensitivity to any received blow, all announce their insensibility.

If I wished to praise sloths and launch a counterattack against Buffon, I could add quite a mouthful at this point. A conventional defense would emphasize neglected features that might inspire human respect. For example, general slowness notwithstanding, sloths can give a quick and nasty slash with those long and inflexible nails that Buffon denigrated (males do fight for usual mammalian reasons of sexual competition; and sloths will defend themselves since they truly can't run away). Moreover, their torpor (and algal cover) do serve an adaptive function in forging inconspicuousness in the presence of enemies, and should not be interpreted as a burden of phyletic primitivity.

I could also point out, still framing a conventional defense by trying to arouse human attention, that sloths have evolved an array of interesting and unique features. For example, sloths are not a dying remnant, but a group in reasonable vigor with more than half a dozen species in two genera—*Bradypus,* the three-toed sloth; and *Choloepus,* the two-toed sloth. With just one or two other exceptions, all mammals have exactly seven cervical (neck) vertebrae (see chapter 16)—yes, even giraffes (where the usual seven are mighty long). But sloths, for some unknown reason, vary this nearly universal number. *Choloepus* has only six cervicals; while *Bradypus* has nine. As a result of these extra vertebrae, *Bradypus* can rotate its head through 270 degrees, or a full three-quarters of a turn!—not quite the full spinning of cartoon clichés (remember Pinocchio turning to display his school clothes to Gepetto), but the closest equivalent in the real world.

Too many sloth lovers, myself included, have tried to stick up for these maligned edentates by invoking such a strategy—that is, by making them either nice or interesting in human terms. Goffart, for example, continues to combat Buffon's calumny two centuries later when he writes:

> Though explorers often described sloths as expressionless, dreamy and stupid, those acquainted with them as pets find that they have a great variety of expressions. Tirler says that when its face is in repose a good-natured smile is forever on its lips. When relieving itself, *Choloepus* has an expression of quiet pleasure.

But the more I ponder the subject, the more I conclude that we should just try to know sloths as *they* perceive and record the world—and not just scan their repertoires for items that resonate with us, or bring us pleasure (including the sublime delight of a good outcome in the outhouse!). And yet, to really know, I need those sixty seconds *within* a *Bradypus* brain—and no power on earth can supply this gift and tool. So I ponder the riddles of ordinary human walking seen as Keystone Kop freneticism, or of reaching for a leaf at t'ai-chi speed as another creature's perception of average pacing. So near into that skull of a distant mammalian relative; so far to know directly.

For my second Costa Rican favorite, I turned to the carrion-feeding raptors, particularly the turkey vulture. I united these birds with the maximally disparate sloths in my mind because both made me wonder so powerfully about "different worlds" in the heads of animals with lifestyles so starkly in contrast with our own choices and proclivities—the only world we can know directly. But I then discovered another connection quite unknown to me at the time. Both these creatures elicited maximal contempt from the greatest arbiter of historical taste—Georges Buffon. I have already quoted Buffon's deprecations of sloths. Now consider his opinion of vultures:

> The eagle attacks his enemies or his victims one on one . . . Vultures, on the other hand, join together in troops, like cowardly assassins, and would rather be robbers than warriors, birds of carnage rather than birds of prey. In this genus [vultures], there are those who gang up upon their prey, several upon one; and there are others intent only upon cadavers, which they rip apart down to the bones. Corruption and infection attract them, instead of repelling them . . . If we compare these birds to mammals, the vulture joins the force and cruelty of tigers with the cowardice and gluttony of jackals, which also unite in troops in order to devour carrion and tear apart cadavers. The eagle, on the other hand, has the courage, the nobility, the magnanimity and the munificence of the lion.

Buffon also tells us that we can easily distinguish vultures from eagles by the naked head and neck of the nasty carrion feeders versus the full feathering of the noble hunters. If this aristocratic French naturalist had known the supposed adaptive value of the vulture's naked head, he would undoubtedly have demoted these birds even further in his estimation—for we now remember Buffon mostly for his celebrated motto *"le style c'est l'homme même"* (the style is the man himself). Vultures plunge their entire head deep into rotting corpses, and a conventional mat of feathers would soon become dangerously fouled, while gore does not adhere to the smooth and naked skin. To cite a standard source (Leslie Brown and Dean Amadon's *Eagles, Hawks, and Falcons of the World*): "Without this denudation, the head feathers would become smeared and matted with gore and infection might occur." (I have little sympathy for adaptationist scenarios in the "just-so story" mode, but this particular tale makes good sense to me, especially since the Old and New World vultures are not closely related by genealogy, but have independently evolved this highly localized loss of feathers—apparently for the same functional reason.)

Little about these birds could possibly be judged as pleasant in human terms. Of the turkey vulture, my Costa Rican source of observation, Buffon concluded in an adjectival frenzy: "They are voracious, cowardly, disgusting, odious and, as with wolves, just as noxious during their lives as they are useless after their death." Consider the grandest of New World vultures, the great (and nearly extinct) California Condor, with maximal wingspan among all the world's flying birds. I don't wish to compromise the noble efforts now under way to save this magnificent species (for adherence to human ethical standards could scarcely be more irrelevant in our judgment of other animals), but descriptions of feeding condors can scarcely inspire any visceral affection.

In the standard source on condor behavior, written in the early 1950s before the population had declined so precipitously, Carl B. Koford describes how a group of condors rips and struggles so vigorously at a carcass that the whole complex (of feeding birds and dead food source) slides slowly downhill:

> Carcasses up to the size of a deer are generally dragged downhill as the condors feed. Once I saw twenty condors feed on a young calf . . . Soon after vigorous feeding commenced the carcass moved

down the slope steadily, attended by several struggling condors, until it was two hundred yards downhill from its original site.

In our current climate of emphasis upon "family values," I won't dwell upon details of their manner of feeding. Suffice to say that the hides of sheep, deer, and cattle (the major sources of larger carcasses) are hard to penetrate—and that condors therefore begin by ripping away at natural orifices, and sticking their smooth heads into the opening bounty.

But I would still give (almost) anything for sixty seconds inside a turkey vulture's head. What does their world look like, as they circle silently above a carcass? What attracts them? What is their aesthetic? Does rot and corruption truly appeal—and, if so, the more the better, or only up to a certain point? Would I, the homunculus in the vulture's brain, view (and smell) a dead cow on the plains as a human explorer might regard a pot of gold at the rainbow's end, or an oasis in the desert?

As these questions emerged in Costa Rica, several thousand miles from my library, I did not realize that an old and substantial literature had developed on this very subject—or at least on the strictly limited and operational way that humans can approach such questions by probing from our restricted position outside the bird's own conceptual world. In particular, naturalists have long wondered and argued about how vultures find their prey.

This old issue immediately raises two questions that both set the puzzle and complexify the answer. First, birds, in general, are preeminently visual animals, particularly so for the raptors (eagles, hawks, and their relatives) that stand in close genealogical proximity to some vultures. But carrion might be found better by smell than by sight. Do vultures therefore use a most unbirdlike sense of smell to find their food? Second, as mentioned before, "vulture" is a functional term for large, carrion-feeding birds that have converged upon a set of common features from different genealogical roots. If we discover that one species can't smell at all, we cannot conclude that another species (with a different evolutionary ancestry) might not use olfaction above all other senses.

The Old World vultures do, apparently, rely entirely upon sight. They take no notice of the most odoriferous parcel of deliciously rotting meat unless they can see the food. But some New World vultures do use smell as a primary sense. Debate has long centered upon the species I saw in Costa Rica, the turkey vulture *Cathartes aura*.

The argument goes back at least to Audubon, who, in 1826, read a technical paper before the Natural History Society of Edinburgh titled "Account of the habits of the turkey buzzard, particularly with the view of exploding the opinion generally entertained of its extraordinary power of smelling." Audubon interpreted his ambiguous experiments as indicating that vultures could not smell, and located prey only with a keen sense of vision. He may have been correct for the species he studied—not the turkey vulture, as he thought and misidentified, but the black vulture, *Coragyps*. The issue therefore remained open for my Costa Rican species.

As criticism of Audubon mounted, his friend, the eminent American naturalist John Bachman, performed a second set of experiments, supposedly to confirm Audubon's conclusion. He even gathered a group of learned and respectable citizens to observe his work and sign a document of assent (shades of Joseph Smith and official witnesses to the Mormon tablets).

Charles Darwin, as a young man in South America on the *Beagle,* took up the subject and, as usual, both asked the key question in the most fruitful way, and got the answer basically right. He did a crude experiment on the Andean condor *(Vultur gryphus)* and concluded that this species does not smell. Darwin wrote:

> Remembering the experiments of Mr. Audubon . . . I tried the following experiment: the condors were tied, each by a rope, in a long row at the bottom of a wall; and having folded up a piece of meat in white paper, I walked backwards and forwards, carrying it in my hand at the distance of about three yards from them, but no notice whatever was taken. I then threw it on the ground, within one yard of an old male bird; he looked at it for a moment

with attention, but then regarded it no more. With a stick I pushed it closer and closer, until at last he touched it with his beak; the paper was then instantly torn off with a fury, and at the same moment, every bird in the long row began struggling and flapping its wings. Under the same circumstances, it would have been quite impossible to have deceived a dog.

But Darwin recognized that other species might smell, and he mentioned that evidence for the turkey vulture favored olfaction as an important sense:

> The evidence in favor of and against the acute smelling powers of carrion-vultures is singularly balanced. Professor Owen has demonstrated that the olfactory nerves of the turkey-buzzard *(Cathartes aura)* are highly developed; and on the evening when Mr. Owen's paper was read at the Zoological Society, it was mentioned by a gentleman that he had seen the carrion-hawks in the West Indies on two occasions collect on the roof of a house, when a corpse had become offensive from not having been buried: in this case, the intelligence could hardly have been acquired by sight.

(I might add that Darwin also developed quite a fondness for Andean condors, despite their dubious lifestyle in human terms. He ended his discussion of this species by writing: "When the condors are wheeling in a flock round and round any spot, their flight is beautiful . . . It is truly wonderful . . . to see so great a bird, hour after hour, without any apparent exertion, wheeling and gliding over mountain and river.")

The issue of olfaction in turkey vultures was not conclusively solved until 1964, when Kenneth E. Stager presented overwhelming evidence, based on years of clever and careful experimentation, that *Cathartes aura* does indeed rely upon a keen sense of smell to find carrion. Turkey vultures will often make an initial identification by sight (though first clues can also be olfactory). They then circle the carcass far above and in a wide arc until they catch a sniff downwind. The famous circle (our conventional icon for vultures) then shortens considerably in radius as the birds home in by odor before descending for the feast.

Ironically, Stager discovered that previous researchers often misread the evidence for olfaction because they assumed "the more putrid the better"—and therefore tested birds with truly rotten meat. In fact, turkey vultures prefer food only slightly rotten and will reject highly putrid flesh if any alternative exists (or unless severe hunger demands compromise with usual standards). Stager writes: "*Cathartes* shows a preference for food that is relatively fresh rather than putrid. If food is in short supply in a given area, the turkey vulture will feed on carrion that is well advanced in putrefaction. Tests . . . indicated that captive *Cathartes* showed a decided preference for recently dead, newly hatched chicks, rather than for putrefied carrion."

I am grateful for all this good information, but I would so prefer my unobtainable minute in a turkey vulture's brain, particularly at the first sight or whiff of a good dead meal on the plains below.

Such conjectures inevitably bring up the contentious theme of animal "consciousness." I confess that I find this subject, as usually debated, both tedious and utterly fruitless as a dispute about subjective *words* pursued by people who mistakenly think that they are arguing about important and resolvable *things.* If I ask "does a dog have consciousness," the endless and passionate arguments that ensue usually reduce to different definitions of this confusing word, rather than focusing upon interesting and empirically resolvable questions about what dogs can and don't do. (I also admit, of course—for this essay takes the point as its major theme—that many questions raised by this fascinating topic do treat the genuine [things rather than words] but unknowable issue of a dog's internal state of mind.) Whether or not a dog "thinks" or "has consciousness" depends upon a chosen definition. Some people won't grant "consciousness" to any creature that can't abstract a general concept—truth or religion, for example—from particulars and then apply the apparatus of formal logic to make inferences even further away from starting points. Others confer "consciousness" upon creatures that recognize kin and remember places of previous danger or pleasure. By the first criterion, dogs don't; by the second, they do. But dogs remain dogs, feeling what they feel without regard to our chosen labels.

In the context of Costa Rica, and international efforts to preserve biodiversity, this issue assumes centrality because the classical argument for why a supposedly decent and moral creature like *Homo sapiens* can mistreat and even extirpate other species rests upon an extreme position in a continuum. The Cartesian tradition, formulated explicitly in the seventeenth century, but developed in "folk" and other versions throughout human history no doubt, holds that other animals are little more than unfeeling machines, with only humans enjoying "consciousness," however defined. Under extreme versions of this theory, even the overt pain and suffering of other mammals (so palpable to us in the most visceral way because the vocal and facial expressions of such close evolutionary relatives match our own reactions to the same stimuli) only record an automatic response with no internal representation in feeling—because other animals have no consciousness. Thus, taking the argument further, we might worry about extinction for other reasons, but not for any aggregate pain or distress in the requisite killing.

I don't think that many people today hold such a strong version of the Cartesian position, but the tradition of viewing "lower" animals as "less feeling" certainly persists as a Band-Aid of justification for our rapacity—just as our racist ancestors argued that "insensitive" Indians couldn't feel conceptual or philosophical pain in the loss of environment or lifestyle (so long as reservations provided bodily needs of food and shelter), and that "primitive" Africans wouldn't lament a forcibly lost land and family so long as slavery provided corporeal security.

I don't want to press the counterargument to extremes. Any definition of consciousness must involve gradations. I am willing to believe that my unobtainable sixty seconds within a sponge or a flatworm might not reveal any mental acuity that I would care to call consciousness. But I am also confident—without wrapping myself in unresolvable arguments about definitions—that vultures and sloths, as close evolutionary relatives with the same basic set of organs, lie on our side of any meaningful (and necessarily fuzzy) border—and that we are therefore not mistaken when we look them in the eye and see a glimmer of emotional and conceptual affinity. I feel sure that I could make something of those sixty seconds if I could ever get in. Vultures must have an aesthetic, and sloths must have a sense of pace.

Modern sloths include but a small remnant of their former diversity—two small-bodied, tree-dwelling genera. As recently as ten to fifteen thousand years ago, giant ground sloths as big as elephants still inhabited the Americas. (Their large and well-preserved skeletons are often mistaken for dinosaurs by people who don't read museum labels.) The other groups of modern edentates were also decimated in the recent past; the giant glyptodonts, better armored than turtles, are fossil armadillos, for example.

South America had been an island continent, far bigger and far more diverse than Australia, for tens of millions of years before the Isthmus of Panama rose just a couple of million years ago. The resulting flood of North American mammals across the new land bridge corresponds in time with the decimation of the native South American fauna (though the causal links remain much disputed). In fact, most large mammals generally considered distinctively South American—jaguars, llamas, and tapirs, for example— are all recent migrants from North America. A few South American forms also managed to move north—including the armadillo of our Southern states, and the so-called (and misnamed) Virginia opossum. But most distinctive South American lineages simply died out—including the borhyaenid marsupial carnivores, the giant and rapacious phorusrhacid ground birds, the horselike (but unrelated) litopterns, and the camel-like (but similarly unrelated and phyletically unique) *Macrauchenia*—see chapter 7. I so wish that this wondrously diverse and evolutionarily disparate fauna had survived— and I do blame the entirely natural rise of the Isthmus of Panama for triggering this particular biological tragedy.

Can We Truly Know Sloth and Rapacity?

Thus, older sloths saw what nature can achieve in her unplanned fortuity. Living species are now experiencing what humans can do with more pressing and conscious power. If I could have those sixty seconds within *Bradypus,* would I not hear a lament for lost and giant brethren; would I not receive a plea for humans to pause, reassess—and, above all, slow down?

2 1

REVERSING ESTABLISHED

ORDERS

WE ALL KNOW HOW THE WORLD WORKS. A FISHERMAN ASKS HIS BOSS IN Shakespeare's *Pericles:* "Master, I marvel how the fishes live in the sea," and receives the evident response, "Why, as men do a-land; the great ones eat up the little ones." Consequently, when humorists invent topsy-turvy worlds, they reverse such established orders and then emphasize the rightness of their absurdity. Alice's Wonderland works on the principle of "sentence first—verdict afterwards." In Gilbert and Sullivan's town of Titipu, the tailor Ko-ko, condemned to death by decapitation, is elevated instead to the rank of Lord High Executioner because—it is so obvious, after all—a man "cannot cut off another's head until he's cut his own off." Pish-tush explains all this in a spirited song with a rousing chorus: "And I am right, and you are right, and all is right too-loora-lay."

Social and literary critics of the so-called postmodernist movement have emphasized, in a cogent and important argument often buried in the impenetrable jargon of their discourse, that conventional support for established orders usually relies upon claims for the naturalness of "dualisms" and "hierarchies." In creating dualisms, we divide a subject into two contrasting categories; in imposing hierarchy upon these dualisms, we judge one category as superior, the other as inferior. We all know the dualistic hierarchies of our social and political lives—from righteous versus infidel of centuries past to the millionaire CEOs who deserve tax cuts versus single mothers who should lose their food stamps in our astoundingly mean-spirited present. The postmodernists correctly argue that such dualisms and hierarchies represent our own constructions for political utility (often nefarious), rather than nature's factual and inevitable dictate. We may choose to parse the world in many other ways with radically different implications.

Our categorizations of nature also tend to favor dualistic hierarchies based upon domination. We often divide the world ecologically into predators and prey, or anatomically into complicated and dominant "higher" animals versus simpler and subservient "lower" forms. I do not deny the utility of such parsings in making predictions that usually work—big fish do generally eat little fish, and not vice versa. But the postmodernist critique should lead us to healthy skepticism, as we scrutinize the complex and socially embedded reasons behind the original formulations of our favored categories. Dualism with dominance may primarily record a human imposition upon nature, rather than a lecture directed to us by the birds and bees.

Natural historians tend to avoid tendentious preaching in this philosophical mode (though I often fall victim to such temptations in these essays). Our favored style of doubting is empirical: if I wish to question your proposed generality, I will search for a counterexample in flesh and blood. Such counterexamples exist in abundance, for they form a staple in a standard genre of writing in natural history—the "wonderment of oddity" or "strange ways of the beaver" tradition. (Sorry to be so disparaging—my own ignoble dualism, I suppose. The stories are terrific. I just often yearn for more intellectual generality and less florid writing.)

Much of our fascination with "strange cases" lies in their abrogation of accepted dualisms based on dominance—the "reversing established orders" of my title. As an obvious example, and paragon of this literature, carnivorous plants have always elicited primal intrigue—and the bigger and more taxonomically "advanced" the prey, the more we feel the weirdness. We yawn when a Venus's-flytrap ensnares a mosquito, but shiver with substantial discomfort when a large pitcher plant devours a bird or rodent.

I keep a file marked "Reversals" to house such cases. I have long been on the lookout for optimal examples, where all three of the most prominent dualisms based on dominance suffer reversal: predator and prey, high and low, large and small—in other words, where a creature from a category usually ranked as small in body, primitive in design, and subject to predation eats another animal from a category generally viewed as bigger, anatomically superior, and rapacious. I now have four intriguing examples, more than enough for an essay. Since we postmodernists abjure hierarchical ranking, I will simply present my stories in the nonjudgmental chronological order of their publication (though postmodernism in this sense—and truly I am not a devotee of this movement—may be a cop-out and an excuse for not devising a better logical structure for this essay!)

1. FROGS AND FLIES. Frogs eat flies. If flies eat frogs, then we might as well be headed for bedlam or the apocalypse. My colleague Tom Eisner of Cornell University is revered throughout our profession as the past master of natural oddities with important and practical general messages. One day in August 1982, at a small pond in Arizona, Eisner and several colleagues noted thousands of spadefoot toads congregating on the muddy shore as they emerged to adulthood in near synchrony from their tadpole stage. Eisner and colleagues described their discovery in a technical publication (see the 1983 article by R. Jackman and others, listed in the bibliography):

> Spaced only centimeters apart in places, they were all of minimal adult size (body length, 1.5 to 2 cm [less than an inch]). Conspicuous

among them were toads that were dead or dying, apparently having been seized by a predator in the mud and drawn partly into the substrate, until only their head, or head and trunk, projected above ground. We counted dozens of such semisubmerged toads.

They then dug deeper and, to their great surprise, found the predator: "a large grublike insect larva, subsequently identified as that of the horsefly *Tabanus punctifer.*" In other words, flies can eat toads! (Although astonishment may be lessened in noting that the tiny toads are much smaller than enormous fly larvae.) Unusually large insects and maximally small vertebrates have also been featured in the few other recorded cases of such reversals—frogs, small birds, even a mouse, consumed by praying mantids, for example.

The fly larvae force themselves into the mud, rear end first, until their front end, bearing the mouthparts, lies flush with the surface. The larvae then catch toads by hooking their pointed mandibles into the hind legs or belly, and then dragging the toad partway into the mud. The larvae—please remember that many tales in natural history are not pleasant by human standards—then suck the toad dry (and dead) by ingesting blood and body fluids only.

I loved the wry last sentence of the paper by Eisner and his colleagues—unusual in style for a technical article, but odd stories have always permitted some literary license:

> The case we report is a reversal of the usual toad-eats-fly paradigm, although . . . the paradigm may also prevail in its conventional form. Adult *Scaphiopus* [the spadefoot toad] might well on occasions have predatory access to the very *Tabanus* flies that as larvae preyed upon their conspecifics.

J. Greenberg, reporting for *Science News* (November 5, 1983), began his commentary with the emotional impact of such reversals:

This is the Okeechobee Fla. Little League team thrashing the New York Yankees; this is Wally Cox beating out Burt Reynolds for the girl; this is Grenada invading the United States. "This is unlike anything I've ever seen," says Thomas Eisner.

2. LOBSTERS AND SNAILS. Decapod crustaceans (lobsters, crabs, shrimp) eat snails, as all naturalists know. In fact, the classic case of an extended evolutionary "arms race," elegantly documented over many years by my colleague Geerat Vermeij, involves increased strength of crab claws correlated with ever more efficient protective devices (spines, ribs, thicker and wavier shells) in snails over geological time. Land crabs are the overwhelmingly predominant predator of my own favorite subject for research, the Caribbean land snail *Cerion*. If snails eat decapods, we might as well retire.

Amos Barkai and Christopher McQuaid studied rock lobsters and whelks (snails of middling size) in waters around two islands, Marcus and Malgas, located just four miles apart in the Saldanha Bay area of South Africa. On Malgas, as all God-fearing folk would only rightly suspect, rock lobsters eat mollusks, mostly mussels and several species of whelks. Barkai and McQuaid write in their 1988 account: "The rock lobsters usually attacked the whelks by chipping away the shell margin with their mouthparts."

The local lobstermen report that, twenty years ago, rock lobsters were equally common on both islands. But lobsters then disappeared from Marcus Island, for unclear reasons, perhaps linked to a period of low oxygen in surrounding waters during the 1970s. In the absence of lobsters as the usual top predator, extensive mussel beds have become established, and the population density of whelks has soared. Barkai and McQuaid asked themselves: "Why do rock lobsters not recolonize Marcus Island despite the high availability of food?"

In an attempt to answer their own question, they performed the obvious experiment—and made an astonishing discovery. The food has become the feeder—this time by overwhelming in number, not equaling in size (the whelks are much smaller than the lobsters). The conventional passive voice of scientific prose does not convey excitement well, but a good story easily transcends such a minor limitation. So, in Barkai and McQuaid's own words, and without any need for further commentary from me (I would only be tempted to make some arch and utterly inappropriate statement about slave revolts—Spartacus and all that):

> One thousand rock lobsters from Malgas Island were tagged and transferred to Marcus Island . . . The result was immediate. The apparently healthy rock lobsters were quickly overwhelmed by large numbers of whelks. Several hundreds were observed being attacked immediately after release and a week later no live rock lobsters could be found at Marcus Island . . . The rock lobsters escaped temporarily by swimming, but each contact with the substratum resulted in several more whelks attaching themselves until weight of numbers prevented escape. On average each rock lobster was killed within fifteen minutes by more than three hundred *Burnupena* [whelks] that removed all the flesh in less than an hour.

Sic semper tyrannis.

3. FISH AND DINOFLAGELLATES. Fish don't generally eat dinoflagellates; why should they even deign to notice such microscopic algae, floating in the plankton? But dinoflagellates certainly don't eat fish; the very notion, given the disparity in sizes, is ludicrous to the point of incomprehensibility.

Dinoflagellates do, however, *kill* fish, by indirect mechanisms long known and well studied for their immense practical significance. Under favorable conditions, dinoflagellate populations can soar to 60 million organisms per liter of water. These so-called blooms can discolor and poison the waters—"red tide" is the most familiar example—leading to massive deaths of fish and other marine organisms.

J. M. Burkholder and a group of her colleagues from North Carolina State University have studied toxic blooms associated with fish kills in estuaries of the southeastern United States. The largest event resulted in the death of nearly one million Atlantic menhaden in the estuary of the Pamlico River. The oddity of this case lies not in the killing of fish per se, a common consequence of dinoflagellate blooms. We have always regarded the deaths of fishes and other marine organisms during red tides as passive and "unintended" results of dinoflagellate toxins, or other consequences of massive algal populations during blooms. No one had supposed that dinoflagellates might actively kill fish as an evolved response for their own explicit advantage, including a potential nutritional benefit for the algal cells. And yet the dinoflagellates do seem to be killing and eating fishes in a manner suggesting active evolution for this most peculiar reversal.

The dinoflagellate lives in a dormant state, lying on the sea floor within a protective cyst. When live fish approach, the cyst breaks and releases a mobile cell that swims, grows, and secretes a powerful, water-soluble neurotoxin, killing the fish. So far, so what?—though the presence of fish does seem to induce activity by the dinoflagellate (breaking of the cyst), thus suggesting a direct link. Anatomical and behavioral evidence both suggest that dinoflagellates have actively evolved their strategy for feeding on fishes. The swimming cell, breaking out from the cyst, grows a projection, called a peduncle, from its lower surface. The cells seem to move actively toward dead or dying fishes. Flecks of tissue, sloughed off from the fish, then become attached to the peduncle and get digested. The authors describe this reversal at maximum disparity in size among my four cases:

> The lethal agent is an excreted neurotoxin. [It] induces neurotoxic signs by fish including sudden sporadic movement, disorientation, lethargy and apparent suffocation followed by death. The alga has not been observed to attack fish directly. It rapidly increases its swimming velocity to reach flecks of sloughed tissue from dying fish, however, using its peduncle to attach to and digest the tissue debris.

4. SPONGES AND ARTHROPODS. Among invertebrates, sponges rank as the lowest of the low (the bottom rung of any evolutionary ladder), while arthropods stand highest of the high (just a little lower than the angels, that is, just before vertebrates on a linear list of rising complexity). Sponges have no discrete organs; they feed by filtering out tiny items of food from water pumped through channels in their body. Arthropods grow eyes, limbs, brains, and digestive systems; many live as active carnivores. Most arthropods wouldn't take much notice of a lowly sponge, but we can scarcely imagine how or why a sponge might subdue and ingest an arthropod.

However, in a 1995 article, crisply titled "Carnivorous Sponges," J. Vacelet and N. Boury-Esnault of the Centre d'Oceanologie of Marseille have found a killer sponge (about as bizarre as a fish-eating dinoflagellate— but both exist). Relatives of this sponge, members of the genus *Asbestopluma*, have only been known from very deep waters (including the all-time record for sponges at more than 25,000 feet), where behavior and food preferences could not be observed. But Vacelet and Boury-Esnault found a new species in a shallow-water Mediterranean cave (less than one hundred feet), where scuba divers can watch directly.

The deep sea is a nutritional desert, and many organisms from such habitats develop special adaptations for procuring large and rare items (while relatives from shallow waters may pursue a plethora of smaller prey). *Asbestopluma* has lost both filtering channels through the body and the specialized cells (called choanocytes) that pump the water through. So how does this deep-water sponge feed?

The new species grows long filaments that extend out from the upper end of the body. A blanket of tiny spicules, or small skeletal projections, covers the surface of the filaments. The authors comment: "The spicule cover . . . gives the filaments a 'Velcro'-like adhesiveness"—the key to this feeding reversal at maximal anatomical distance for invertebrates. The sponge captures small crustaceans on the filaments—and they can't escape any more than a fuzz ball can detach itself from the Velcro lining of your coat pocket.

The authors continue: "New, thin filaments grew over the prey, which was completely enveloped after one day and digested within a few days." The sponge, in other words, has become a carnivore.

Four fascinating stories to give us pause about our preconceptions, particularly our dualistic taxonomies based on the domination of one category over another. The little guys sometimes turn tables and prevail—often enough, perhaps, to call the categories themselves into question.

I see another message in these reversals—a consequence of the reassessment that must always proceed when established orders crumble, or merely lose their claim to invariance. In our struggle to understand the history of life, we must learn where to place the boundary between contingent and unpredictable events that occur but once and the more repeatable, lawlike phenomena that may pervade life's history as generalities. (In my own view of life, the domain of contingency looms vastly larger than all Western tradition, and most psychological hope, would allow. Fortuity pervades the origin of any particular species or lineage. *Homo sapiens* is a contingent twig, not a predictable result of ineluctably rising complexity during evolution—see the end of chapter 15 for Darwin's view on this issue.)

The domain of lawlike generality includes broad phenomena not specific to the history of particular lineages. The ecological structure of communities should provide a promising searching ground, for some principles of structural organization must transcend the particular organisms that happen to occupy a given role at any moment. I imagine, for example, that all balanced ecosystems must sustain more biomass as prey than as predators—and I would accept such statements as predictable generalities, despite my affection for contingency. I would also have been willing to embrace the invariance of other rules for sensible repetition—that single-celled creatures don't kill and eat large multicellular organisms, for example. But these four cases of reversed order give me pause.

In a famous passage from the *Origin of Species,* Charles Darwin extolled the invariance of certain ecological patterns by using observed repetition in independent colonizations to argue against a range of contingently unpredictable outcomes:

> When we look at the plants and bushes clothing an entangled bank, we are tempted to attribute their proportional numbers and kinds to what we call chance. But how false a view is this! Every one has heard that when an American forest is cut down, a very different vegetation springs up; but it has been observed that the trees now growing on the ancient Indian mounds, in the Southern United States, display the same beautiful diversity and proportion of kinds as in the surrounding virgin forests. What a struggle between the several kinds of trees must here have gone on during long centuries, each annually scattering its seeds by the thousand; what war between insect and insect—between insects, snails, and other animals with birds and beasts of prey—all striving to increase, and all feeding on each other or on the trees or their seeds and seedlings, or on the other plants which first clothed the ground and thus checked the growth of the trees! Throw up a handful of feathers, and all must fall to the ground according to definite laws; but how simple is this problem compared to the action and reaction of the innumerable plants and animals which have determined, in the course of centuries, the proportional numbers and kinds of trees now growing on the old Indian ruins!

But the same patterns do not always recur from adjacent starting points colonized by the same set of species. Even the most apparently predictable patterns of supposedly established orders may fail. Remove the lobsters from waters around one South African island, and a new equilibrium may quickly emerge—one that actively excludes lobsters by converting their former prey into a ganging posse of predators!

Thus, I sense a challenge in these four cases, a message perhaps deeper than the raw peculiarity of their phenomenology—and the resulting attack upon our dualistic and hierarchical categories. We do not yet know the rules of composition for ecosystems. We do not even know if rules exist in the usual sense. I am tempted, therefore, to close with the famous words that D'Arcy Thompson wrote to signify our ignorance of the microscopic world (*Growth and Form,* 1942 edition). We are not quite so uninformed about the rules of composition for ecosystems, but what a stark challenge and what an inspiration to go forth: "We have come to the edge of a world of which we have no experience, and where all our preconceptions must be recast."

BIBLIOGRAPHY

Bahn, P. G. and J. Vertut. 1988. *Images of the Ice Age.* New York: Facts on File.

Barber, L. 1980. *The Heyday of Natural History.* Garden City, N.Y.: Doubleday.

Barkai, A., and C. McQuaid. 1988. "Predator-Prey Role Reversal in a Marine Benthic Ecosystem." *Science* 242: 62–64.

Barnosky, A. 1985. "Taphonomy and Herd Structure of the Extinct Irish Elk *Megaloceras giganteus.*" *Science* 228: 340–44.

Barnosky, A. 1986. "The Great Horned Giants of Ireland: The Irish Elk." *Carnegie Magazine* 58: 22–29.

Boyle, R. 1661. *The Sceptical Chymist.* London: J. Cadwell.

Boyle, R. 1688. *A Disquisition About the Final Causes of Natural Things.* London: H. C. for John Taylor.

Brace, C. L. 1977. *Human Evolution.* 2nd ed. New York: Macmillan.

Brace, C. L. 1991. *The Stages of Human Evolution.* Englewood Cliffs, N.J.: Prentice Hall.

Breuil, H. 1906. *L'Evolution de la peinture et de la gravure sur murailles dans les cavernes ornées de l'age du renne.* Paris: Congres Prehistorique de France.

Breuil, H. 1952. *Les Figures Incisées et Ponctuées de la Grotte de Kiantapo.* Belgium: Musée Royal du Congo Belge.

Brooks, W. K. 1889. "The Lucayan Indians." *Popular Science Monthly* 36: 88–98.

Brooks, W. K. 1889. "On the Lucayan Indians." *Memoirs of the National Academy of Sciences* 4: 2, 213–23.

Brown, L., and D. Amadon. 1968. *Eagles, Hawks, and Falcons of the World.* New York: McGraw-Hill.

Bibliography

Buffon, G. 1752. *Histoire Naturelle*. Paris.

Burkholder, J. M., E. J. Noga, C. H. Hobbs, and H. B. Glassgow Jr. 1992. "New 'Phantom' Dinoflagellate Is the Causative Agent of Major Estuarine Fish Kills." *Nature* 358: 407–10.

Chambers, R. 1853. *A Biographical Dictionary of Eminent Scotsmen*. Glasgow: Blackie.

Chauvet, J. 1996. *Chauvet Cave: The Discovery of the World's Oldest Paintings*. London: Thames and Hudson.

Clutton-Brock, T. 1982. "The Functions of Antlers." *Behavior* 79:108–25.

Coe, M. D. 1992. *Breaking the Maya Code*. New York: Thames and Hudson.

Cuvier, G. 1812. *Recherches sur les ossemens fossiles*. Paris: Deterville.

Dagg, A., and J. B. Foster. 1976. *The Giraffe: Its Biology, Behavior, and Ecology*. New York: Van Nostrand Reinhold Co.

Dana, J. D. 1852–55. *Crustacea*. Philadelphia: C. Sherman.

Dana, J. D. 1857. *Thoughts on Species*. Philadelphia.

Dana, J. D. 1863. "On Parallel Relations of the Classes of Vertebrates, and on Some Characteristics of the Reptilian Birds." *American Journal of Science* 36:315–21.

Dana, J. D. 1863–76. "The Classification of Animals Based on the Principle of Cephalization." *American Journal of Science,* 1863, 36:321–52, 440–41; 1864, 37:10–33, 157–83; 1866, 41:163–74; 1876, 12:245–51.

Dana, J. D. 1872. *Corals and Coral Islands*. New York: Dodd & Mead.

Dana, J. D. 1876. *Manual of Geology*. 2nd ed. New York: Ivision, Blakeman, Taylor and Company.

Darwin, C. 1842. *The Structure and Distribution of Coral Reefs*. London.

Darwin, C. 1851–54. "A Monograph on the Fossil Cirripedes of Great Britain. (Lepadidae, Balanidae, Verrucidae)." London: Palaeontographical Society.

Darwin, C. 1859. *On the Origin of Species*. London: John Murray.

Darwin, C. 1868. *The Variation of Animals and Plants Under Domestication*. London: John Murray.

Darwin, C. 1881. *The Formation of Vegetable Mould Through the Action of Worms*. London: John Murray.

De Robertis, E. M., and Y. Sasai. 1996. "A Common Plan for Dorsoventral Patterning in Bilateria." *Nature* 380: 37–40.

Bibliography

Delage, Y. 1884. *Evolution de la Sacculine* (Sacculina Carcini *Thomps*): *Crustace endoparasite de l'ordre nouveau des Kentrogonides.* Paris: Centre National de la Recherche Scientifique.

Dickens, C. 1859 1895. *All the Year Round.* London: Chapman and Hall.

Dickens, C. 1865. *Our Mutual Friend.* Philadelphia: T. B. Peterson & Brothers.

Dimery, N. J., R. McN. Alexander and K. A. Deyst. 1985. "Mechanics of the Ligamentum Nuchae of Some Artiodactyls." *Journal of Zoology* 206: 341–51.

Du Chaillu, P. B. 1861. *Explorations and Adventures in Equatorial Africa.* London: John Murray.

Egerton, J. 1995. *Turner: The Fighting* Temeraire. London: National Gallery Publications.

Farago, C. 1996. *Leonardo da Vinci: Codex Leicester: A Masterpiece of Science.* New York: American Museum of Natural History.

Figuier, L. 1863. *La Terre Avant la Deluge.* Paris: Hachette.

François, V. and E. Bier. 1995. "The *Xenopus chordin* and the *Drosophila short gastrulation* Genes Encode Homologous Proteins Functioning in Dorsal-Ventral Axis Formation." *Cell* 80: 19–20.

François, V., M. Solloway, J. W. O'Neill, H. Emery, and E. Bier. 1994. "Dorsal-Ventral Patterning of the *Drosophila* Embryo Depends on a Putative Negative Growth Factor Encoded by the *short gastrulation* Gene. *Genes and Development* 8: 2602–26.

Gaskell, W. H. 1908. *The Origin of Vertebrates.* London: Longmans, Green and Co.

Geoffroy St. Hilaire, E. 1822. "Considérations générales sur la vertèbre." Paris, *Memoires du Museum National D'Histoire Naturelle* 9: 89–119.

Gerace, D. T., ed. 1987. "Columbus and His World." San Salvador, Bahamian Field Station. Fort Lauderdale: The Station.

Glenner, H., and J. T. Høeg. 1995. "A New Motile, Multicellular Stage Involved in Host Invasion by Parasitic Barnacles (Rhizocephala)." *Nature* 377: 147–50.

Goffart, M. 1971. *Function and Form in the Sloth.* New York: Pergamon Press.

Gosse, P. H. 1856. *The Aquarium: An Unveiling of the Wonders of the Deep Sea.* 2nd ed. London: J. Van Voorst.

Gould, S. J. 1965. "Is Uniformitarianism Necessary?" *American Journal of Science* 263: 223–28.

Gould, S. J. 1974. "The Origin and Function of 'Bizarre' Structures: Antler Size and Skull Size in the 'Irish Elk.'" *Evolution* 28: 191–220.

Bibliography

Gould, S. J. 1977. *Ontogeny and Phylogeny.* Cambridge, Mass.: Belknap Press of Harvard University Press.

Gould, S. J. 1986. "Knight Takes Bishop?" *Natural History* 95:5, 18–33.

Gould, S. J. 1992. "Red in Tooth and Claw." *Natural History* 101:11, 14–23.

Gould, S. J. 1994. "Lucy on the Earth in Stasis." *Natural History* 103:9, 12–20.

Greenberg, J. 1983. "Poetic Justice in the Arizona Desert." *Science News* 124:19, 293.

Gross, C. G. 1993. "Hippocampus Minor and Man's Place on Nature: A Case Study in the Social Construction of Neuroanatomy." *Hippocampus* 3: 403–13.

Hibberd, S. 1858. *Rustic Adornments,* 2nd ed. London: Groombridge and Sons.

Hitching, F. 1983. *The Neck of the Giraffe: Darwin, Evolution, and the New Biology.* New York: New American Library.

Høeg, J. T. 1985. "Cypris Settlement, Kentrogon Formation and Host Invasion in the Parasitic Barnacle *Lernaeodiscus porcellanae* (Muller) (Crustacea:Cirripedia: Rhizocephala)." *Acta Zoologica* 66: 1–45.

Holley, S. A., P. D. Jackson, Y. Sasai, B. Lu, E. M. De Robertis, F. M. Hoffmann, and E. L. Ferguson. 1995. "A Conserved System for Dorsal-Ventral Patterning in Insects and Vertebrates Involving *Sog* and *Chordin*." *Nature* 376: 249–53.

Huxley, T. H. 1863. *Evidence as to Man's Place in Nature.* New York: D. Appleton.

Jackman, R., S. Nowicki, D. J. Aneshanslex, and T. Eisner. 1983. "Predatory Capture of Toads by Fly Larvae." *Science* 222: 515–16.

James, H. 1898. *The Two Magics: The Turn of the Screw.* London: Macmillan & Co.

Johanson, D. C. and D. Edgar. 1996. *From Lucy to Language.* London: Weidenfeld & Nicolson.

Kemp, M. 1981. *Leonardo da Vinci: The Marvellous Works of Nature and Man.* Cambridge, Mass: Harvard University Press.

Kennedy, D. H. 1983. *Little Sparrow: A Portrait of Sophia Kovalevsky.* Athens: Ohio University Press.

Kimmel, C. B. 1996. "Was Urbilateria Segmented?" *Trends in Genetics* 12: 329–31.

Kingsley, C. 1863. *The Water Babies.* New York: Macmillan.

Kipling, R. 1902. *Just So Stories for Little Children.* New York: Doubleday.

Kircher, A. 1664. *Mundus Subterraneus.* Amsterdam: J. Jansson.

Kitchener, A. 1987. "Fighting Behaviour of the Extinct Irish Elk." *Modern Geology* 11: 1–28.

Bibliography

Koford, C. B. 1953. *The California Condor.* New York: National Audubon Society.

Kovalevsky, V. 1980. *The Complete Works of Vladimir Kovalevsky.* Edited by S. J. Gould. New York: Arno Press. (Contains French articles on horses, 1873; German article on horses, 1876; English article on artiodactyls, 1874; and two German articles on artiodactyls, 1876.)

Krafft-Ebing, R. von. 1892. *Psychopathia Sexualis.* London: F. A. Davis Co.

Lamarck, J. B. 1809. *Philosophie Zoologique.* Paris: Dentu.

Lankester, E. R. 1880. *Degeneration: A Chapter in Darwinism.* London: Macmillan and Co.

Leroi-Gourhan, A. 1967. *Treasures of Prehistoric Art.* New York: H. N. Abrams.

Levin, I. 1976. *The Boys from Brazil.* New York: Random House.

Linnaeus, C. 1758. *Systema Naturae.* Stockholm.

Linnaeus, C. 1759. *Genera Morborum.* Upsala.

Linnaeus, C. 1771. *Fundamenta Testaceologiae.* Upsala: ex officena Edmanniana.

Lister, A. M. 1994. "The Evolution of the Giant Deer, *Megaloceros giganteus* (Blumembach)." *Journal of the Linnean Society of London* 112:65–100.

Lowell, P. 1906. *Mars and Its Canals.* New York: Macmillan Company.

Lutzen, J. 1984. "Growth, Reproduction, and Life Span in *Sacculina carcini* Thompson (Cirripedia:Rhizocephala) in the Isefjord, Denmark." *Sarsia* 69: 91–106.

Lyell, C. 1832–33. *Principles of Geology.* London: J. Murray.

MacCurdy, E. 1939. *The Notebooks of Leonardo da Vinci.* New York: Reynal & Hitchcock.

Mayr, E. 1963. *Animal Species and Evolution.* Cambridge, Mass.: Belknap Press of Harvard University Press.

McKay, D. S., E. K. Gibson Jr., K. L. Thomas-Keprta, H. Vali, C. S. Romanek, S. J. Clemett, X. D. F. Chillier, C. R. Maechling, and R. N. Zare. 1996. "Search for Past Life on Mars: Possible Relic Biogenic Activity in Martian Meteorite ALH84001." *Science* 273: 924–30.

Mendes da Costa, E. 1757. *A Natural History of Fossils.* London: L. Davis and C. Reymers.

Mendes da Costa, E. 1776. *Elements of Conchology, or, An Introduction to the Knowledge of Shells.* London: B. White.

Mitchell, G. F., and H. M. Parkes. 1949. "The Giant Deer in Ireland." *Proceedings of the Royal Irish Academy* 52(B)(7):291–314.

Bibliography

Mivart, St. George J. 1871. *On the Genesis of Species.* New York: D. Appleton and Co.

Morison, S. E. 1942. *Admiral of the Ocean Sea.* Boston: Little, Brown and Co.

Muller, F. 1864. *Für Darwin.* Leipzig: Wilhelm Engelmann.

Nichols, J. 1817–58. *Illustrations of the literary history of the eighteenth century. Consisting of authentic memoirs and original letters of eminent persons; and intended as a sequel to the Literary anecdotes.* London: Nichols, Son, and Bentley.

Osborn, H. F. 1918. *The Origin and Evolution of Life: On the Theory of Action, Reaction and Interaction of Energy.* London: G. Bell.

Owen, R. 1846. *A History of British Fossil Mammals, and Birds.* London: J. Van Voorst.

Owen, R. 1865. *Memoir on the Gorilla.* London: Taylor and Francis.

Owen, R. 1866. *Memoir of the Dodo.* London: Taylor and Francis.

Paley, W. 1802. *Natural Theology.* London: R. Faulder.

Patten, E. 1920. *The Grand Strategy of Evolution.* Boston: R. G. Badger.

Rehbock, P. F. 1980. "The Victorian Aquarium in Ecological and Social Perspective." In M. Sears and D. Merriman, eds. *Oceanography: The Past, International Congress on the History of Oceanography.* New York: Springer Verlag.

Richter, P. J. 1883. *The Literary Works of Leonardo da Vinci.* London: S. Low, Marston, Searle & Rivington.

Ritchie, L. E., and J. T. Høeg. 1981. "The Life History of *Lernaeodiscus porcellanae* (Cirripedia:Rhizocephala) and Co-Evolution with its Porcellanid Host." *Journal of Crustacean Biology* 1: 334–47.

Rolfe, W. D. Ian, 1983. "William Hunter (1718–1783) on Irish 'Elk' and Stubbs's Moose." *Archives of Natural History* 11: 263–90.

Rudwick, M. J. S. 1992. *Scenes from Deep Time.* Chicago: University of Chicago Press.

Rupke, N. A. 1994. *Richard Owen: Victorian Naturalist.* New Haven: Yale University Press.

Ruspoli, M. 1987. *The Cave of Lascaux: The Final Photographs.* New York: Abrams.

Sauer, C. O. 1966. *The Early Spanish Main.* Berkeley: University of California Press.

Scheuchzer, J. J. 1732–37. *Physique Sacrée, ou Histoire-Naturelle de la Bible.* Amsterdam: Chez P. Schenk.

Smith, G. 1968. *The Dictionary of National Biography.* Oxford: Oxford University Press.

Stager, K. E. 1964. "The Role of Olfaction in Food Location by the Turkey Vulture *(Cathartes aura)." Los Angeles County Museum Contributions in Science* 81:63.

Bibliography

Strickland, H. E. and A. G. Melville. 1848. *The Dodo and Its Kindred.* London: Reeve, Benham, and Reeve.

Stringer, C. 1996. *African Exodus.* London: Cape.

Swisher III, C. C., W. J. Rink, S. C. Anton, H. P. Schwarcz, G. H. Curtis, A. Suprijo, and Widiasmoro. 1996. "Latest *Homo erectus* of Java." *Science* 274:5294, 1870–74.

Thompson, D. W. 1942. *On Growth and Form.* Cambridge: Cambridge University Press.

Turner, R. 1992. *Inventing Leonardo.* Berkeley: University of California Press.

Tyson, E. 1699. *Orang-outang, sive,* Homo sylvestris, *or the Anatomy of a Pygmie compared with that of a Monkey, an Ape, and a Man.* London: T. Bennet, Daniel Brown, and Mr. Hunt.

Vacelet, J., and N. Boury-Esnault. 1995. "Carnivorous Sponges." *Nature* 373. 333–35.

Wallerius, J. G. 1747. *Mineralogia, Eller Mineralriket, Indelt och beskrifvit.* Stockholm.

Whitehead, P. J. P. 1977. "Emmanuel Mendes da Costa (1717–1791) and the *Conchology, or Natural History of Shells." Bulletin of the British Museum of Natural History* (Historical Series), 6: 1–24.

Wood, J. G. 1868. *The Fresh and Salt-Water Aquarium.* London: G. Routledge.

ILLUSTRATION CREDITS

Grateful acknowledgment is made for permission to reproduce the following:

page 32 *La Gioconda* (the Mona Lisa), Leonardo da Vinci, Louvre, Paris, France. Neg. no. 93DE1846. Photograph by RMN–R. G. Ojeda.

pages 37, 39, 40, 41 Sketches by Leonardo da Vinci from the Leicester Codex. Courtesy of Seth Joel/Corbis Corporation.

page 49 *The Fighting* Temeraire *Tugged to Her Last Berth to Be Broken Up, 1838,* J.M.W. Turner, 1839, copyright © National Gallery, London.

pages 66, 68 From *Physica Sacra*, J. J. Scheuchzer, 1730s. Photographs by Jackie Beckett, American Museum of Natural History.

pages 71, 72 From *The Earth Before the Flood*, Louis Figuier, 1863, and 4th ed., 1865. Photographs by Jackie Beckett, American Museum of Natural History.

page 73 From *Rustic Adornments*, Shirley Hibberd, 2nd ed., 1858. Photographs by Jackie Beckett, American Museum of Natural History.

page 79 From *Fundamenta Testaceologiae*, Carolus Linnaeus, 1771.

page 127 From *On the Classification and Geological Distribution of the Mammalia,* Richard Owen, 1859.

page 151 From monograph, Vladimir Kovalevsky, 1876. Photograph by Craig Chesek, American Museum of Natural History.

page 164 Lascaux Cave painting. Photograph by Jean-Marie Chauvet/Sygma. Courtesy of the French Government Tourist Office.

page 183 *The Moose*, George Stubbs, 1770, Hunterian Art Gallery, University of Glasgow, Scotland.

page 185 *Megaloceros* from Cougnac Cave, southwest France. Modified after Lorblanchet et al., 1993, by A. M. Lister, 1994, *Zoological Journal of the Linnean Society,* London.

Illustration Credits

page 191   Skeleton of an Irish elk, Richard Owen, 1846. Neg. no. 2A23132. Courtesy of the Department of Library Services, American Museum of Natural History.

page 207   Copyright © 1997 Donald C. Johanson. Used by permission of Nevraumont Publishing Co., New York.

page 219   Photographs courtesy of Sally Walker.

page 221   Photograph by Jackie Beckett, American Museum of Natural History.

page 236   From *Memoir of the Dodo,* Richard Owen, 1866. Neg. no. 5848. Courtesy of the Department of Library Services, American Museum of Natural History.

page 238   LEFT: From *Memoir of the Dodo*, Richard Owen, 1866. Neg. no. 5847. RIGHT: From *The Dodo and Its Kindred,* H. E. Strickland and A. G. Melville, 1848. Neg. no. 5869. Both courtesy of the Department of Library Services, American Museum of Natural History.

page 248   Illustration by John Tenniel from *Alice's Adventures in Wonderland*, Lewis Carroll, 1865. Copyright © Corbis-Bettmann.

page 253   Engraving of Anton Von Werner's original painting, copyright © The Granger Collection, New York.

page 292   *Evolution*, copyright © Vint Lawrence, 1995.

page 307   *La giraffe,* Georges Buffon.

page 322   From *The Origin of Vertebrates,* Walter Holbrook Gaskell, 1908.

pages 324, 325   From *Memoires du Museum d'Histoire Naturelle,* Geoffroy Saint-Hilaire, 1822. Neg. nos. 338680, 337423. Photographs by Jackie Beckett. Courtesy of the Department of Library Services, American Museum of Natural History.

page 333   E. M. De Robertis and Y. Sasai; reprinted with permission from *Nature,* vol. 380. Copyright © 1996 Macmillan Magazines Limited.

page 341   From *Mars and Its Canals,* Percival Lowell (New York: Macmillan, 1906). Courtesy of the Lowell Observatory, Arizona.

page 354   Copyright © UPI/Corbis-Bettmann.

page 357   From *Degeneration: A Chapter in Darwinism,* E. Ray Lankester (London: Macmillan and Co., 1880).

page 362   Courtesy of Archives de Zoologie Expérimentale Deuxième Série, Tome II, 1884.

page 365   Beth Beyerholm; with permission from *Nature,* vol. 377. Copyright © 1995 Macmillan Magazines Limited.

page 368   Courtesy of Nancy J. Haver, from *Invertebrates,* R. C. Brusca and G. J. Brusca (Sinauer Associates, 1990).

page 397   *Reverse Evolution* copyright © Vint Lawrence, 1995.

INDEX

Index

Index

Index

Index

Index

Index

Index

Index